FLOOD INUNDATION MODELING AND HAZARD MAPPING UNDER UNCERTAINTY IN THE SUNGAI JOHOR BASIN, MALAYSIA

ANUAR BIN MD. ALI

FLOOD INUNDATION MODELING AND HAZARD MAPPING UNDER UNCERTAINTY IN THE SUNGAI JOHOR BASIN, MALAYSIA

DISSERTATION

Submitted in fulfilment of the requirement of
the Board for Doctorates of Delft University of Technology
and of the Academic Board of the IHE Delft
Institute for Water Education
for
the Degree of DOCTOR
to be defended in public
on Monday, March 26, 2018 at 15:00 hours
in Delft, the Netherlands

by

Anuar bin Md. ALI

Master of Science in Water Engineering
Universiti Putra Malaysia, Serdang, Malaysia

born in Kg. Jawa, Kota Tinggi, Johor, Malaysia

This dissertation has been approved by the supervisors:
Prof. dr. D.P. Solomatine
Prof. dr. G. Di Baldassarre

Composition of Doctoral Committee:

Chairman	Rector Magnificus TU Delft
Vice-Chairman	Rector IHE Delft
Prof. dr. D.P. Solomatine	IHE Delft / TU Delft, promotor
Prof. dr. G. Di Baldassarre	Uppsala University, Sweden, promotor

Independent members:
Dr. M. N. B. Md. Noh	Department of Irrigation and Drainage, Malaysia
Prof. dr. J.C.J.H. Aerts	Vrije Universiteit Amsterdam
Prof. dr. S. Grimaldi	Tuscia University, Italy
Prof. dr. W.G.M. Bastiaanssen	IHE Delft / TU Delft
Prof. dr. ir. H.H.G. Savenije	TU Delft, reserve member

CRC Press/Balkema is an imprint of the Taylor & Francis Group, an informa business

Published by:
CRC Press/Balkema
Schipholweg 107C, 2316 XC, Leiden, the Netherlands
Pub.NL@taylorandfrancis.com
www.crcpress.com – www.taylorandfrancis.com
ISBN 978-1-138-60334-9

This thesis is dedicated to my late parents.
Who born me, raised me, supported me, taught me and loved me.

Allahyarhamah Hajjah Halijah Bt Md. Yassin
(06 August 1946 ~ 06 April 2015)
Allahyarham Haji Md. Ali Bin Hj. Ismail
(31 December 1935 ~ 11 June 2016)

Al-Fatihah

Summary

Flood is a natural disaster that occurs almost regularly in Malaysia particularly during the monsoon seasons. Hence, it is of no surprise that flood is considered one of the most significant natural hazards in the country in terms of number of affected population, fatalities and economic damage. One of the efforts to minimize flood losses is providing useful information through floodplain inundation maps, i.e. spatial distribution of flood hazard. Traditionally, many modellers have used deterministic approaches in flood inundation modelling. Deterministic approaches are based on a single simulation with the "best fit model" and do not explicitly consider uncertainties in model parameters, terrain data, and model structure. When model results are then used to generate a flood hazard map, neglecting uncertainties may lead to precise, but inaccurate maps and lead to wrong or misleading information to decision makers. Thus, the scientific literature has recently proposed a number of probabilistic methods to recognize, assess and account for uncertainties affecting flood inundation modelling. In this context, this research work aims to contribute to this research work by further exploring the impact of various sources of uncertainty on the results of hydraulic models. The case study of this research is the Sungai Johor river basin in Malaysia. Both 1-D and 2-D hydraulic models were utilized.

In using 1-D hydraulic models, the geometric description of rivers and floodplains is performed by using a number of cross-sections, which play an important role in the accuracy of model results. In this work, criteria for cross-section spacing were tested and verified via numerical experiments.

Similarly, digital elevation models (DEMs) used as geometrical input significantly affect the results of flood inundation modelling exercises. DEM is essential input that provides topographical data in flood inundation modelling. However, it can be derived from several sources either through remote sensing techniques (space-borne or air-borne imagery) or from more traditional ground survey methods. These DEMs are characterized by different precision and accuracy. This study quantified the effect of using different DEM data source and resolution in a 1-D hydraulic modelling of floods.

This study also explored the differences arising from the use of deterministic and uncertainty approaches in deriving design flood profiles and flood inundation maps. To this end, the generalized likelihood uncertainty estimation (GLUE) technique was used and the uncertainty in model predictions was derived through Monte Carlo analysis. In particular, this work focused on impact of uncertain inflow data and roughness coefficients in the accuracy of flood inundation models.

As part of this research, 2-D hydraulic modelling software (LISFLOOD-FP) was also used to assess the effect of spatial data re-sampling (e.g. from high to low resolution) on model outcomes. This study evaluated two re-sampling techniques with combination of three different aggregation functions, i.e. minimum, maximum and mean values.

This research work has not only provided useful results, but has also suggested further research and improvement of flood risk and mapping practices. The knowledge generated by, as well as the findings of this thesis, will be transferred to other study areas in Malaysia.

Samenvatting

Overstromingen komen regelmatig voor in Maleisië, met name tijdens de Monsoon. Het is daarom niet verrassend dat in Maleisië overstromingen worden gerekend tot de belangrijkste natuurrampen als het gaat om het aantal getroffen personen, dodelijke slachtoffers en economische schade. Een van de inspanningen die kan worden gedaan om schade door overstromingen te beperken is het verstrekken van betekenisvolle informatie in de vorm van overstromingsgevaarkaarten die de overstromingsgevoelige gebieden weergeven. Tot op heden hebben veel modelleurs de traditionele deterministische benadering gebruikt voor overstromingsmodellen. De deterministische benadering is gebaseerd op één simulatie met het "best fit model" (het meest nauwkeurig geachte model). Onzekerheden in parameterwaarden, het onderliggend hoogtemodel en modelstructuur worden niet expliciet meegenomen. Als deze modelresultaten dan worden gebruikt om een overstromingsgevaarkaart te genereren, leidt het verwaarlozen van onzekerheden tot een precieze, maar mogelijk onnauwkeurige kaart en daarmee tot foutieve of misleidende beslissingsondersteunende informatie. De recente wetenschappelijke literatuur heeft daarom een aantal probabilistische methoden voorgesteld om onzekerheden bij overstromingsmodellering te erkennen, in te schatten, en weer te geven. In deze context beoogt dit onderzoek bij te dragen door de invloed van verschillende onzekerheden op de uitkomsten van hydraulische modellen verder te verkennen. Het stroomgebied van de Sungai Johor rivier in Maleisië dient als praktijkvoorbeeld. Zowel 1-D als 2-D hydraulische modellen werden toegepast.

Bij 1-D hydraulische modellen bepalen dwarsprofielen, die de geometrie van rivierbedding en overstromingsvlakte beschrijven, in hoge mate de nauwkeurigheid van modelresultaten. In dit onderzoek werden numerieke experimenten uitgevoerd om verschillende criteria voor het vaststellen van de afstand tussen twee opeenvolgende dwarsprofielen in het model te evalueren.

Op vergelijkbare wijze bepalen digitale hoogtemodellen (DEMs), wanneer die worden gebruikt als geometrische input, in hoge mate de resultaten van overstromingsmodellen. Een DEM levert essentiële topografische input voor het modelleren van overstromingen. Een DEM kan echter worden afgeleid van verschillende brongegevens, zoals van Remote Sensing technieken enerzijds (vanuit lucht of ruimte) of van traditionele landmetingen anderzijds. De resulterende DEMs verschillen in precisie en nauwkeurigheid. Dit onderzoek kwantificeerde de invloed van het gebruik van verschillende brongegevens en resolutie voor de DEMs op de 1-D modellering van overstromingen.

Dit onderzoek analyseerde ook de optredende verschillen als gevolg van het gebruik van deterministische en probabilistische methoden voor het bepalen van maatgevende overstromingsprofielen en overstromingskaarten. Hiertoe is de 'Generalised Likelihood Uncertainty Estimation' (GLUE) techniek gebruikt en de onzekerheid in modelvoorspellingen is bepaald met de 'Monte Carlo' techniek. De analyse spitste zich toe op de invloed van onzekerheden in de watertoevoergegevens en ruwheidsfactoren op de nauwkeurigheid van overstromingsmodellen.

Als onderdeel van dit onderzoek werd ook 2-D hydraulische modelleringssoftware (LISFLOOD-FP) gebruikt, om het effect van 're-sampling' van ruimtelijk data (bijvoorbeeld van hoge naar lage resolutie) op modelresultaten te bepalen. Dit onderzoek heeft twee 're-sampling' technieken geëvalueerd in combinatie met drie verschillende aggregatiefuncties; minimum, maximum, en gemiddelde waarde.

Dit onderzoek heeft niet alleen nuttige resultaten opgeleverd, maar ook suggesties voor vervolgonderzoek en verbetering van overstromingsrisicokartering in de praktijk. De gegenereerde kennis, zowel als de bevindingen van deze dissertatie, zullen worden meegenomen bij andere praktijkstudies in Maleisië.

This abstract is translated from English to Dutch by Dr. Schalk Jan Van Andel, Senior Lecturer, Integrated Water Systems and Governance Department, IHE Delft.

TABLE OF CONTENTS

Chapter 1
Introduction

1.1 Background

Flooding is the most significant natural hazard in Malaysia in terms of number of affected population, fatalities and economic damage. Since 1920, the country has experienced major flood events in 1926, 1963, 1965, 1967, 1969, 1971, 1973, 1979, 1983, 1988, 1993, 1998, and 2005 and most recently in December 2006 and January 2007 which occurred in Johor. According to the Emergency Events Database (EM-DAT), more than 300 flood disasters were reported in Malaysia between 1960 and 2009. It was estimated that these flood events affected more than 1 million people and caused some 300 fatalities.

Figure 1.1: Flood prone areas in Malaysia (adapted from DID, 2003)

According to DID (2003), the total flood prone area in Malaysia is around 30,000 km², while the country total area is 328,799 km² (see Figure 1.1). It is also estimated that as 2000, 22% of the total population of Malaysia, which counts 22.2 million people, lives in this flood prone area. In term of economic damage, as at 2000, the total Annual

Average Flood Damage for Malaysia is estimated €212.0 million, compared to €23.0 million in 1980.

Floods are relatively common in Malaysia because of the geographical characteristic of the country that gets an abundance of rainfall during the monsoon season in addition to convection storms during the hot but humid periods. In Malaysia, rivers and their floodplains fulfil a variety of functions for both human use and natural ecosystems. Yet, 85 out of 189 main river basins are characterised by frequent and damaging flood events (DID, 2009).

1.2   Problem statement

Johor is the fifth largest and one of the most developed state in Malaysia with an area of 19,210 km² and population of about 3 million. The recent severe flooding (December 2006 to January 2007) that occurred in the state caught many in surprise in terms of its magnitude, extent as well as the huge resultant damages amounting to more than €350.0 million for public facilities alone according to a preliminary estimate. Sungai Johor river basin, which is the largest basin in Johor State with a total area of approximately 2,690 km², was one of the worst hit areas, being affected by two major floods within a short period of time.

During the December 2006 and January 2007 floods, Sungai Johor and all its major tributaries overtopped the banks and cause massive flooding throughout the catchments. 14,864 flood victims from 3,303 families and 15,660 victims from 3,483 families were evacuated in Kota Tinggi during first wave (19-24 December 2006) and second wave (11-17 January 2007), respectively (*source:* www.jkmnj.gov.my). The first event occurred from 19th to 24th December 2006, during which widespread rainfall of about 350 mm was recorded. The second flood event occurred only three weeks later, from 11th to 17th January 2007, with a widespread heavy rainfall of 400mm (3-day total). The worst hit settlement areas included Kota Tinggi town. The inundation depth in Kota Tinggi town was about 3 m.

In response, the Department of Irrigation and Drainage of Malaysia (DID) made an initiative to carry out a technical study in the Johor state that covered all river basins including the Sungai Johor basin. The main scope of work for the technical study was to identify the appropriate solutions to minimize impact of floods in each affected area. The proposed solutions were not limited to structural methods but also comprised of non structural elements, one of which was to develop flood hazard map. The flood hazard map was intended to be one of the main basis to formulate appropriate flood management plan to assist the Authorities in handling any possible flood events in the future.

The flood hazard map developed for the Sungai Johor basin helps in assist in the assessment and management of flood risk, however not all uncertainties associated with this problem were considered, not all available data sources were used (like LiDAR), and the models used could be better fine-tuned. There are a number of ways to improve flood hazard mapping for Sungai Johor basin, and it is our intention to do it in this study.

1.3 Flood mapping

Flood mapping is an issue addressed in many countries. It is worth noting that the European Union (EU) has adopted a new directive known as EU Floods Directive (EU, 2007) that proposed a transition from traditional flood defence approaches to holistic flood risk management strategies (Di Baldassarre *et al.*, 2009). The main objective of the European directive is to reduce and manage flood risk by implementing comprehensive management plans, which include flood hazard and inundation maps to be prepared by all the Council Members by 2013.

Merz et al. (2007) noted that flood maps are effective tools for assisting flood hazard management. The requirement and classification of flood maps depends on the purpose of their use. Flood hazard maps, in particular, can be defined as maps showing inundated area or different parameters such as flood depth and flood velocity.

Several important parameters are required for performing hydraulic flood modelling such as topographic data, discharge data to provide model inflow and outflow as boundary conditions, estimation of the roughness coefficient and validation data (Bates 2004).

A substantial of research have been made to investigate the flood hazards, not only to understand the behaviour of flood flow (i.e. in river channel and floodplain), but also the characteristics of flood such as occurrences, magnitude and extent. Most of this effort was reasonably carried out by conducting the hydraulic modelling of floods (Horrit and Bates, 2002; Patro *et al.*, 2009; Di Baldassarre and Montanari, 2009; Poretti and De Amicis, 2011; Crispino *et al.*, 2014). Furthermore, the output from hydraulic modelling of floods for instance in estimation of inundations area and flood profile is useful information's for assisting the decision makers in flood relief planning and operations.

Although maps of flood hazard provide useful indication on the potentially inundated area and negative impact posed by flood, there is significant uncertainty associated with these maps (Di Baldassarre *et al.*, 2009). Unfortunately, although modellers are well aware that significant approximation affects flood hazard assessment and various methods to deal with uncertainty have been recently developed, the awareness among environment and river basins agencies, authorities and engineering consultancies is still lacking as the advances in uncertainty analysis are hardly applied. To facilitate a wider application of these methods, the development of clear methods is therefore needed (Di Baldassarre *et al.,* 2010).

1.4 Uncertainty in flood hazard mapping

The most common representation of of flood inundation modelling results remains a deterministic approach based on a single simulation using a best fit model. Unfortunately, this approach does not explicitly account for the uncertainties in the modelling process (Bates *et al.,* 2004) and may lead to a precise but inaccurate hazard assessment (Di Baldassarre *et al.,* 2010), despite increasing knowledge in flood propagation and inundation processes.

Although ample literature has been discussed to identified the source of uncertainties in flood inundation mapping (Bales and Wagner, 2009; Domeneghetti *et al.,* 2013; Dottori *et al.,* 2013; Jung *et al.,* 2013; 2014), but to eliminate the uncertainties completely are impossible due to various limitations such as computational times, cost, technology and knowledge of the flood science itself.

Uncertainty in flood hazard mapping may arise from accuracy of topographic data (i.e. source of data sets), topographic data types (TIN, Raster/GRID), precision (cell size/resolution), spacing of river cross-sections, model parameter (e.g. Manning's n roughness coefficient) or hydraulic modelling approach (i.e. 1-D, 2-D/3-D). For instance, the accuracy and precision of the topographic data sets used in extracting the cross sections for a hydraulic model and the mapping of water surface elevations may affects estimation of the flood hazard area in term of area and depth. Besides that, integrating a data from different format may also add another uncertainty in flood hazard mapping. As an example, it's common to integrate between surveyed river cross-section data with existing topographic floodplain data.

1.5 Research questions

The proposed study aims to address the following research questions:

i.   How do many sources of uncertainty (e.g. hydrologic data, topographic data, and model selection) affect flood hazard mapping?

ii.   What are the potentials and limitations of different data sources (including remote sensing) in supporting flood inundation modelling?

iii.   How can we model uncertainty to better define safety levels in the design of flood protection structures?

1.6 Aim and research objective

The general aim of this study is to develop a model-based methodological framework allowing for flood mapping and thus assisting public administrator in making appropriate decisions under uncertainty, with application to Sungai Johor basin.

The specific objectives of this research are as follows:

i.   To identify the most relevant sources of uncertainty associated with the generation/development of flood hazard maps.

ii.   To develop and integrate the necessary models and data sources (including remote sensing data) allowing for accurate description and prediction of the natural processes leading to flooding, and thus supporting flood mapping.

iii.   To identify the source/effect of uncertainty related to safety levels of flood protection structures.

1.7   Dissertation Structure

This thesis is organised in eight chapters. The first three chapters are general. Chapter 1 provides an overview of the research with concise explanations of its relevance. Research questions and objectives of this thesis are listed and briefly explained. Chapter 2 summarizes a literature review which covers several topics related to the thesis. Chapter 3 highlighted a detailed description of the study area of this thesis: the Johor River which located in Johor River Basin, Malaysia. It also described the data available and used in this study.

Chapter 4 and Chapter 5 addresses the first research question. In particular, Chapter 4 describe the application of different cross-section spacing in 1-D hydraulic flood modelling for the purpose of understanding how its influence to the model output does. Which Chapter 5 demonstrated the applicability of the 2-D hydraulic flood modelling to simulate flood inundation output using the different resolution of DEMs which built from different techniques of re-sampling.

Chapter 6 addresses the research question number two by using different sources of DEMs (with different resolutions) and remote sensing data in 1-D hydraulic modelling.

Chapter 7 addresses the third research question by comparing deterministic and probabilistic approaches for floodplain mapping using 1-D hydraulic modelling.

Lastly, Chapter 8 summaries the findings and presents the conclusion and recommendations.

Chapter 2
Literature Review

2.1 What is floods

Flood is a natural hazard that resulted from combination of hydrological and meteorological factors. It occurs when a normally dry land areas are temporary inundated due to overflowing of water at the natural or artificial confines of a river, including groundwater caused by prolonged or heavy rainfall.(Wisner *et al.,* 2004; Martini and Loat, 2007; Klijn 2009). Hydrologists define flood as a sudden increase in water discharge that caused a sudden peak in the water level. Once flood is over, the water level will drop back to near-constant base flow or no flow. As summarized by Martini and Loat (2007), flooding is when water and/or sediments exist at unwanted areas other than the water body. Whereas, Ward (1978), defined flood as a body of water which is not normally submerged.

2.2 Types of flood

Flood can be categorized into different types based on location of occurrence and what cause them. The major ones are as described below.

River flood

River flood occurs when a river basin is filled with too much water that is more than the capacity of the river channel. River flood is considered as an expected event as it usually occurs seasonally, normally during rainy seasons. The surplus water overflows the river banks and runs into adjoining low-lying lands.

Coastal flood

Flood that occurs in coastal area due to the drive of the ocean waters inland is known as coastal flood. Natural phenomenon such as tropical storm, hurricane or intense offshore low pressure can cause unusually high amount of the ocean water to be driven towards the land resulting in the coastal flood. Similarly, tidal sea waves that happen due to earthquake or volcanic activities in the sea can also caused coastal flood.

Urban flood

Heavy rainfall and changes in the runoff behaviours are the most common reasons for urban floods. The changes in the runoff behaviours is mostly due to the development of the land to buildings and paved roads which have less absorbing ability compared to an undeveloped area or natural fields. The rainfall runoff in the urban areas can be as high as six times than that in a natural fields. As a result, roads become rapid rivers and basements as death traps when they are filled with water.

Flash flood

Flash floods occur when a large amount of water flood within short period of time. Normally it occurs locally and suddenly without or with little warning. Flash floods could happen due to immoderate rainfall or a sudden release of water from a dam.

This research will focus and discuss on river flood and the extent of the flood to the adjacent area along the river.

2.3 Flood prone areas

The areas adjacent to a river prone to flooding can be defined as floodplain and floodway. A flood area that is deep with high flow velocities with presence of debris flow that can cause possible erosion is identified as floodway. There should be no development allowed to take place within the floodway area except for critically necessary infrastructure such as bridges (UNISDR 2002).

A floodplain on the other hand represents the areas surrounding the river channel (including floodway) that can be inundated during the occurrence of a flood (FEMA 2008). The boundary of a floodplain cannot be defined as the magnitude of a flood is limitless. The higher a point in the floodplain is, the lesser the probability of inundation. A flood line however can be drawn up to define a floodplain area based on the water level of a flood with specified annual exceedance probability. No development should take place within this flood line.

2.4 Hazard and flood hazard

It is important to understand and be accustomed with the terms and terminology used in disaster management. However, there are different definitions and terminologies used implicated in term of hazard and flood hazard. Below are the defining term of hazard and flood hazard.

2.4.1 Definition

Hazard

Hazard as defined by the ISDR (2009) is a dangerous phenomenon, human activity or process that may cause loss of life, injury or other health problems, loss or damage of property, livelihoods, infrastructure and services, social and economic disruption or environmental degradation.

Samuels et al. (2009) defined hazard as a physical event, phenomenon or human activity with the potential but not necessarily lead to harm.

Flood hazard

Flood is one of the most commonly occurred environmental hazards that may not necessarily caused by natural events but can also be due to or aggravated by human activities such as deforestation, pollution or uncontrolled urbanization that changes or disrupts the natural landscape.

According to ISDR, only a few hazards, such as earthquakes and hurricanes, are true natural hazards. Flood is categorized as a socio-natural or unnatural hazards where a naturally original disaster aggravated by human factors (ISDR 2009).

In general, flood hazard is the result from a combination of physical exposure represented by the type of flood and their statistical pattern at a particular site, and human vulnerability to geophysical processes. Human vulnerability is associated with keys socio economy such as the numbers of people at risk on the floodplain and the ability of the population to anticipate and cope with the hazard.

Merz et al. (2007) defined flood hazard as the exceedance probability of a potentially damaging flood event in a particular area within a specific period of time. However, this statement does not represent the consequences of such floods to community, environment or development.

A flood hazard statements should taken into account the depth of the process that goes beyond a flood frequency curve such as the inundation depth, flow velocity, duration of the flood occurrence and the rate of water increase, since the consequences of flood not only rely on the intensity of the flood.

As Caddis et al. (2012) highlighted, the definition of flood hazard involve consideration of a various factor such as magnitude of floods, duration of flooding, effective warning time, depth and velocity of floodwaters, flood readiness, evacuation and access and type of development. Most of this factor are quantifiable either from flood modelling (e.g. magnitude, velocity, depth, duration of flood) or through assessment (e.g. land use, roads, human behaviour).

2.5 Flood modelling

Flood modelling is a simplification of the real situation event. A flood model of a particular river basin for example simulates the real flood events that have occurred using the actual hydrological input data, the basin's hydraulic characteristics and boundary conditions. These modelling are able to show effects on the results based on different boundary conditions or input data. Hence by simulation, the behaviour of the flood risk or hydraulic characteristics at a certain period of time can be determined and investigated.

In the development of flood mapping, with recent advances in technology whereby computation time has been tremendously reduced, it is becoming necessary to simulate flood inundations in the flood plains caused by different magnitudes of flood events. Nowadays, different types of inundation models exist and approaches have been made by various researchers by using various hydrodynamic modelling models (Bates *et al.*, 2003). One of the most important developed tools for hydraulic modelling is geographical information system or GIS that allows one or two dimensional representation of computed hydraulic parameters.

Variety of software has been used widely for dynamic 1-D flow simulation in rivers such as MIKE 11, HEC RAS, SOBEK-1-D etc. Even though the 1-D models are simple to use and provide information on bulk flow characteristics, it is however fail to provide information particularly on the flow field. A 2-D model whereas require substantial computer time to provide the information.

As there is a limitation of using 1-D or 2-D numerical models, attempt have been made to couple 1-D river flow models and 2-D floodplain flow models. The coupled between two numerical models offer a great advantage for real time simulation of flood events. Among that coupled models known is SOBEK 1-D-2-D developed by Delft Hydraulics, while Danish Hydraulics Institute (DHI) developed MIKE FLOOD which combination of MIKE 21 and MIKE 11.

2.5.1 Mathematical model application

HEC-RAS modelling

HEC-RAS is a modelling program developed by the US Army Corps of Engineers. It allows two different approaches to be adopted, ie (i) steady flow calculations, and (ii) unsteady flow simulation. The unsteady flow simulation has been used in this study to simulate the flood inundation.

HEC-RAS modelling package uses the 1-D St Venant equation to calculate open channel flow. In the unsteady flow simulation the horizontal exchange of water between channel and floodplain was assumed to be insignificant, and the water discharge is distributed according to the conveyance.

The flow in the channel can be presented as:

$$Q_c = \phi Q \qquad\qquad\qquad (2.1)$$

___

where Q_c is flow in the channel and Q is total flow. Here, ϕ determines how flow is partitioned between the floodplain and channel, based on the conveyance of K_c and K_f. Where ϕ is calculated as

$$\phi = \frac{K_c}{K_c + K_f} \tag{2.2}$$

while K_c is represents as conveyance in the channel and K_f is conveyance in the floodplain. Conveyance is defined as

$$K = \frac{A^{5/3}}{nP^{2/3}} \tag{2.3}$$

where P is wetted perimeter, A is cross-section area and n represents Manning's n roughness coefficient. From the above equation, the 1-D equation can be written as follows:

$$\frac{\partial A}{\partial t} + \frac{\partial \phi Q}{\partial x_c} + \frac{\partial (1 - \phi)Q}{\partial x_f} = 0 \tag{2.4}$$

$$\frac{\partial Q}{\partial t} + \frac{\partial}{\partial x_c}\left(\frac{\phi^2 Q^2}{A_c}\right) + \frac{\partial}{\partial x_f}\left(\frac{(1 - \phi)^2 Q^2}{A_f}\right) +$$
$$gA_c\left(\frac{\partial z}{\partial x_c} + S_c\right) + gA_f\left(\frac{\partial z}{\partial x_f} + S_f\right) = 0 \tag{2.5}$$

where

$$S_c = \frac{\phi^2 Q^2 n_c^2}{R_c^{4/3} A_c^2} \text{ and } S_f = \frac{(1 - \phi)^2 Q^2 n_f^2}{R_f^{4/3} A_f^2} \tag{2.6}$$

where A_c and A_f is the cross sectional area of the flow of the channel and floodplain, x_c and x_f are the distances along the channel and floodplain, R is hydraulic radius

15

(A/P) and *S* is the friction slope. The finite difference method was utilized for discretion of the equations 2.4 and 2.5 and solved using a four-point implicit method.

LisFlood-FP

The LISFLOOD-Floodplain or also known as LISFLOOD-FP is a hydraulic model originally developed by Bates and De Roo (2000). This model has been broadly tested and compared with other models in determine the standard of the model (Neal *et al.*, 2012). Furthermore, the stability of the original numerical solver by Bates et al. 2010 for low friction condition has been improvised by de Almeida et al. (2012). This LISFLOOD-FP works on a 2-D regular grid structure and simulates water flow by solving the shallow water equations in 1-D, without the convective acceleration terms from the momentum equations (Bates *et al.*, 2010).

To calculate the flow, Q between cells, equation 2.7 is used:

$$Q = \frac{q - gh_{flow}\Delta t \frac{\Delta(h+z)}{\Delta x}}{(1 + gh_{flow}\Delta tn^2 |q|/h_{flow}^{10/3})}\Delta x \qquad (2.7)$$

Where *q* is the flux between cells from previous iteration, *g* is gravity, *h_flow* is the maximum depth of flow between cells, *Δt* is the model time-step, *h* is the water depth in each cell, *z* is elevation, *Δx* is the cell width and *n* is a friction coefficient.

Having established the discharge across all four boundaries of a cell, the cell water depth (h) is updated using equation 2.8:

$$\frac{\Delta h^{i,j}}{\Delta t} = \frac{Q_x^{i-1,j} - Q_x^{i,j} + Q_y^{i,j-1} - Q_y^{i,j}}{\Delta x^2} \qquad (2.8)$$

Where cell are indexed in two-dimensions using i and j. To enhance the model robustness, the time step, t which is controlled by shallow water Courant-Friedrich-Levy (CFL) condition was introduced in the LISFLOOD-FP formulation:

$$\Delta t_{max} = \alpha \frac{\Delta x}{\sqrt{gh}} \qquad\qquad (2.9)$$

where α is a coefficient typically defined between 0.3 and 0.7 (Bates *et al.*, 2010).

2.5.2 GIS environment

A hydraulic model is intended to represents the flood physical processes over time of a river channel or flood plain as realistic as possible that able to provide acceptably accurate output for different scenarios to its user (Pullar and Springer, 2000). With GIS, a hydraulic model is presented in a spatial or geographical manner that would allow the model to analyze, predict and solve engineering problems in a more powerful and comprehensive way.

Many modelling application uses GIS as the database manager and visualization tools through the use of Windows Graphical User Interfaces (GUIs) making the output easier to understand by its users. The benefits of GIS integrated modelling are tremendous.

With the integration of GIS in these modelling, some of the techniques or procedures from the manual flood hazard mapping processes may need to be modified or changed, among others include search method, governing algorithms, data requirements and flood inundation extent and depth. (Noman *et al.*, 2001). In GIS, data can be extracted, combined with others or reformatted if needed for various modelling processes and even used to generate other inputs as required by the models (Robbins and Phipps, 1996).

It is important as suggested by Noman *et al.*, (2001) that the integration of GIS in a hydraulic model should be made in such a way to allow automatic data transfer without jeopardizing the ability to replace the hydraulic model with the alternative ones.

The data exchange system between hydraulic model and GIS software was first developed by Evans (1998) using HEC-RAS as the study package. The system enable HEC-RAS to import cross-section coordinates from a terrain model to develop channel and reach geometry and exports the data back to a GIS upon completion of the hydraulic calculations for comparison with the terrain model. In 1998, ESRI further translated and improved Evans' code and with some added utilities enhance its use. The result was an ArcviewGIS extension called AVRas. In general Arcview GIS allows user to work with maps and geographic information.

Study by Tate et al. (1999) to improve the HEC-RAS model's accuracy led to the development of Avenue scripts for Arcview GIS that incorporates data such field survey, stream geometry and control structures into a GIS-based terrain model. A very accurate digital orthography was used to develop the terrain model in this study.

Using this Avenue scripts Merwade et al. (2008), applied GIS techniques to create from linear cross-sections a continuous river bathymetry in the form of a 3D mesh, and integrated this bathymetry with the floodplain topography using a simple smoothing algorithms.

2.5.3   Input data for flood modelling

The performance of any model can only be as good as the data it uses to parameterize it and to calibrate and validate. While models should be selected based on the characteristics of the problem in hand, it is also clear that models of different complexity have different data requirements, and in practice this may constrain user's choice in model selection.

Asselman et al. (2009), highlighted the data required by any hydraulic model are in principal the boundary condition, initial condition, topography data, friction data and hydraulic data for use in model validation. Whereas, Methods et al. (2007) noted the data required for flood modelling are hydraulic boundary determinations, geometric data, discharge data, roughness data and calibration and validation.

In general, modelling of floodplain flooding requires high quality input data, which should include rainfall, a digital terrain model, land use and calibration. Rainfall data should ideally be provided by a dense network of rain gauges and/or the weather radar. Both sources are important, the former is generally considered as more accurate, whilst the latter typically has higher spatial resolution, which enables advanced applications such as nowcasting (quick precipitation forecasting).

Accurate digital terrain model (DTM) is essential for the simulation of floodplain flooding. It is used for sub-catchment delineation, creation of surface flow paths and ponds and as a basis for 2-D modelling. Terrain data that contains information about buildings, walls, kerbs and other surface features is called Digital Elevation Model (DEM).

The followings are the available techniques in obtaining a DEM for a flood modelling : i. aerial stereo-photogrammetry (Baltsavias, 1999; Westaway *et al.*, 2003), ii. airborne laser altimetry or LiDAR and iii. airborne Synthetic Aperture Radar interferometry (Hodgson *et al.*, 2003) and iv. Radar interferometry from sensors mounted on space-borne platforms, in particular the Shuttle Radar Topography Mission (SRTM) data (Rabus *et al.*, 2003).

Currently, LiDAR is the most used techniques in the hydraulic modelling literature (Marks and Bates, 2000; French, 2003; Charlton *et al.*, 2003; Cook and Merwade, 2009).

Land-use data is used to automatically parameterise variables such as roofed and other impervious areas, surface roughness, etc. Land-use images can be provided by remote sensing. Where this technology is not available, key features such as streets, car parks, housing, green areas, etc. can be distinguished from the existing maps and the corresponding imperviousness and roughness can be assigned to each surface type.

Finally, calibration is required in floodplain models to match the flood extent and the flood depth. It is usually difficult to properly calibrate, verify and validate a flooding model due to a lack of comprehensive sets of observed data. Observed flood extents such as flood mark, photos or video record can be invaluable for calibration.

2.6 Uncertainty in flood modelling and mapping

2.6.1 Definition of uncertainty

In recent years, flood disasters have contributed to the realization that the future is in inherently uncertain. In flood risk management, one of the crucial issues is how to deal with the uncertainty. Generally, uncertainties are associated with human behavior, organizations and social system which make it more difficult to predict future vulnerability of area to flooding. Uncertainty reduces the strength of confidence in the estimated cause and effect chain.

Different authors give different interpretation and definition of uncertainty and other related terms such as error, risk and ignorance. Walker et al. (2003) defined uncertainty as the deviation from the ideal complete determinism of knowledge of a relevant system that is not achievable. Pappenberger et al. (2005), meanwhile described uncertainty in a more general concept that reflects the lack of sureness about something that can be as little as just a short of complete sureness to an almost completely lack of conviction about the results. Whereas Refsgaard et al. (2007) on the other hand describes uncertainty as the result of the given information being incomplete or blurred, inaccurate, unreliable or inconclusive, or potentially falsely judge that led a person to be uncertain or lacks confidence about the specific outcomes of an event.

2.6.2 Types of uncertainty

There are many ways to differentiate uncertainties. Apel et al. (2004) and Merz and Thieken (2005, 2009), classified uncertainty into two types known as aleatory uncertainty and epistemic uncertainty (see Table 2.1).

From the point of view of model-based decision support, Walker et al. (2003) distinguish uncertainty into three dimensions as follows.

o The location within the model where the uncertainty shows itself ;

o The level where the uncertainty manifests itself along the spectrum of different levels of knowledge between determinism and total ignorance in terms of statistical uncertainty, scenario uncertainty and recognised uncertainty;

o Nature of uncertainty whether due to the lack or imperfection of knowledge or the inherent variability of the case being described in the study.

Table 2.1: Types of uncertainty (after Apel *et al.*, 2004)

Types of Uncertainty	Description	Other Term
Aleatory uncertainty	Quantities that are inherently variable over time, space, or populations of individuals or object	• Variability, • Objective uncertainty, • Stochastic uncertainty, • Stochastic variability, • Inherent variability, • Randomness, and • Type-A uncertainty.
Epistemic uncertainty	Incomplete knowledge of the object of investigation and is related to our ability to understand, measure, and describe the system under study.	• Subjective uncertainty, • Lack of knowledge/limited knowledge uncertainty, • Ignorance, • Specification error, and • Type-B uncertainty

2.6.3 Sources of uncertainty

Prinos et al. 2008 have identified the likely sources of uncertainties for each element with the variable, and divided it into three types, known as model uncertainty, parameter uncertainty and data uncertainty. Table 2.2 displays the type and likely sources of uncertainties and its variable.

Table 2.2: Sources and types of uncertainty (after Prinos *et al.*, 2008)

Type of uncertainty	Source of uncertainty	Variable			
		A.D	L.F	W.S	F.D
Model uncertainty	Rainfall runoff modelling	*			
	Wave modelling	*			
	Selection of distribution function	*			
	Breach modelling		*		
	Model selection (1-D or 2-D model)			*	
	Steady or unsteady calculation			*	
	Frictional resistance equation			*	
	Dependence on water stage				*
Parameter Uncertainty	Channel roughness			*	
	Channel geometry			*	
	Levee parameters (geometry, substrate, breach width, turf)		*		
	Parameters of the statistical distribution	*			
Data uncertainty	Short or unavailable records	*			
	Measurement errors	*			
	Measurement errors of levee geometry		*		
	Sediment transport and bed forms			*	
	Debris accumulation and ice effects			*	
	Land/building use, value and location				*
	Content value				*
	Structure first-floor elevation				*
	Flood warning time				*
	Public response to a flood				*
	Performance of the flood protection system				*

Notes: *A.D.: Annual Damage, L.F.: Levee Failure, W.S.: Water Stage, F.D.: Flood Damage*

Nevertheless, the uncertainties in flood risk management such as in simulation modelling are principally due to natural variability and knowledge uncertainty. In flood risk mapping, source of uncertainties arising from several factors such as model approach (Cobby *et al.*, 2003; Horritt and Bates, 2002; Horritt *et al.*, 2006; Tayefi *et al.*, 2007), topography (Casas *et al.*, 2006), friction coefficient (Aronica *et al.*, 2002); grid cell size (Werner, 2001), or flow characteristic (Purvis *et al.*, 2008).

2.7 Flood mapping

2.7.1 Types and content of flood map

In the past, government agencies implemented engineering solutions such as dams, levees, seawalls and others in the attempt to reduce flood damage to the communities. However, these solutions often did not reduce flood damage costs and property loss, nor discourage continued development within the flood-prone area.

Now, there is an action to transform the management of flood from a conventional flood defence solutions to a flood risk management approach. In Europe, the European Parliament has adopted a new Flood Directive with the main objective is to establish a framework to assess and manage flood risk (EU, 2007). One of the directive tasks is to produce flood hazards maps and risk maps in every state that will form the basis of a flood risk management plans in the future. Thus, to achieve this directive, flood mapping has become a priority and an important aspect for the EU members.

In the field of flood risk management, the confusion is not only arising in use of risk related definition, but also in the naming of different flood maps (de Moel *et al.*, 2009). For instance, Merz et al. (2007), proposed four type of flood map namely as flood danger map, flood hazard map, flood vulnerability map and flood damage risk map (see Table 2.3). In general, flood map can be defined as a map presents the area prone to flooding at one or more floods with given return periods.

Table 2.3: Proposal for systematic flood mapping at the local scale (Merz *et al.*, 2007)

Type of map	Definition
Flood danger map	Shows the spatial distribution of the flood danger without information about the exceedance probability.
Flood hazard map	Shows the spatial distribution of the flood hazard, i.e. information on flood intensity and probability of occurrence for single or several flood scenarios.
Flood vulnerability map	Shows the spatial distribution of the flood vulnerability, i.e. information about the exposure and/or the susceptibility of flood-prone elements (population, built environment, natural environment).
Flood damage risk map	Shows the spatial distribution of the flood damage risk, i.e. the expected damage for single or several events with a certain exceedance probability.

While, Di Baldassarre et al. (2010) and Merz et al. (2007), classified flood maps into three type of map known as flood hazard map, flood vulnerability maps and flood risk maps. The most common categories of map used to illustrate flood hazard is a map show inundation area for events with different return periods. Table 2.4 show each type of flood map with different content of information.

Table 2.4: Content of flood map (Di Baldassarre *et al.,* 2010)

Type of flood map	Content
Flood hazard map	Displaying intensity of floods and their associated exceedence probability.
Flood vulnerability map	Illustrate the consequences of floods on economy, society and the natural environment.
Flood risk map	Showing the spatial distribution of the risk, which, for natural disasters, can be defined as the probability that a given event will occur multiplied by its consequences.

Nevertheless, as mentioned in European Flood Directive, developments of flood hazard are as one of the directive tasks. van Alphen et al. (2009) have summarized the content, purpose and user of flood hazard map as shown in Table 2.5.

Table 2.5: Content, purpose and users of flood hazard and flood risk maps (modified from van Alphen *et al.*, 2009)

	Flood hazard map
Content	•Flood extent according to probability classes, according to past events. •Flood depth. •Flow velocity. •Flood propagation. •Degree of danger.
Purpose and use	•Land-use planning and land management Watershed management. •Water management planning. •Hazard assessment on local level. •Emergency planning and management. •Planning of technical measures. •Overall awareness building.
Target group/use	•National, regional or local land-use planning. •Flood managers. •Emergency services. •Forest services (watershed management). •Public.

2.7.2 Use of flood maps

Flood maps are created by various institutions and used by many stakeholders. The main producers of flood map either at local scale or basin scale are governmental institution and private company particularly to insurance company or cooperation between the government and private company. van Alphen and Passchier (2007) highlighted that the use of flood maps serve at least one of the three purposes of flood risk management as follow:

- o Preventing the build-up of new risks (planning and construction),
- o Reducing existing risks, and
- o Adapting to changes in risks factors.

Depending on the purpose, the stakeholders may have very specific demands on content, scale, accuracy or readability of the map. van Alphen et al. (2009) noted that the types of flood maps may depend on the user/target group and their interest of using flood maps (see Table 2.6). While, de Moel et al. (2009) divided the use of flood map to three main institutions, flood map use by governmental agency especially for land use planning and emergency purposes (evacuation), private sector mainly in insurance industry and transnational river basin authorities like ICPR (Rhine) and ICPDR (Danube).

Table 2.6: The user/target group and the purpose use of the flood maps (van Alphen *et al.*, 2009)

User/target group	Purpose use of flood map
Land-use planners	To identify the location of flood prone areas and potential inundation depth (to prescribe or advise building codes).
Flood managers	To identify areas with high potential damage and casualties, in order to prioritize their protective efforts.
Emergency services	To identify the locations with large concentrations and numbers of (vulnerable) population, expected flooding patterns, dangerous locations and road availability for evacuation and transport.
Utilities companies	To identify areas with potential disruption of their services and how to minimize this.
Insurance companies	To identify areas in potential damage and related probability.
Communities	To know whether they live in a flood prone area and what their options are in case of flooding.

2.7.3 Flood hazard map in Malaysia

In Malaysia, the flood hazard map has been developed through Department of Irrigation and Drainage, Malaysia (DID). Based on the DID standards, the flood hazard maps shall be produced based on 10, 50, 100 and 200 years ARI at the scale of 1:25,000 with considering the present and future land use conditions. For the specified ARI, the flood hazard maps must clearly indicate the flood depth and flood extent with different colour scheme. However, there is no classification of hazardous of flood either in term of flood depth or flood extent.

For this study, to characterize the site into hazardous classes, the inundation water depths used with modifications after the standards of Japan International Corporation Agency (JICA) were adopted to generate flood hazard maps (see Table 2.7). These water depth classes were based on human characteristics in conformance with the Japanese Flood Fighting Act established in 2001 (DID, 2003).

Table 2.7: Flood hazard classes (adopted from Japanese Flood Fighting Act 2001)

Water depth (m)	Level of hazard	Description of hazard
0 to 0.5	Low	Most houses will stay dry and it is still possible to walk through the water.
0.5 to 1.0	Medium	There will be at least 50 cm of water on the ground floor and electricity will have failed by now.
1.0 to 2.0	High	The ground floor of the houses will be flooded and the inhabitants have either to move to the first floor or evacuate.
More than 2.0	Extreme	The first floor and often also the roof will be covered by water. Evacuation at this stage is the logic choice of action.

Chapter 3
Study area and data availability

3.1 Study area

3.1.1 Administrative

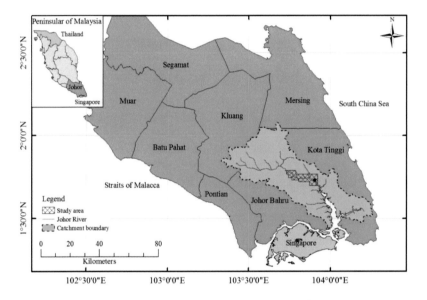

Figure 3.1: Map showing the districts under state of Johor, Malaysia

Johor River is situated in the State of Johor in the southern part of Peninsular of Malaysia at latitudes of 1º 40' and 2º 00' N, longitudes 103º 20' and 104º 00' E. Administratively, the State of Johor is divided into eight districts in which most of Johor River is located entirely within the Kota Tinggi district (see Figure 3.1).

The administrative town of this district is also known as Kota Tinggi and has a total administration area of 3,500 km². 65% of its border is surrounded by the sea. As shown in Figure 3.2, it is recognized that this town area is one of the highly urbanized areas located along Johor River. In term of inhabitants, Kota Tinggi population is increasing from 114,267 people in 1980 to 212,558 in 2005. It projected by year 2020, the population in this district will be increase nearly to 250,000 people. Kota Tinggi is still dependent of primary resources (mainly agriculture such as palm oil plantation, rubber plantation and crops) as the main economic drivers. Alongside agriculture, tourism, mining, timber and industrial are among contributor to the economy of Kota Tinggi.

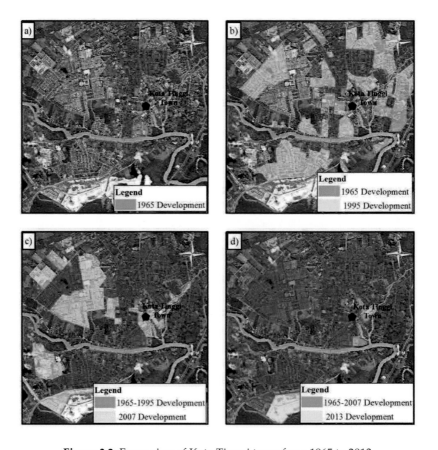

Figure 3.2: Expansion of Kota Tinggi town from 1965 to 2013

3.1.2 River systems

Johor River Basin covers an area of 2,690 km^2 (DID, 2009). The river systems are starting from the Mount Belumut (east of Kluang) and Mount Gemuruh (to the north), towards downstream at Tanjung Belungkor. Johor river is originated from it source of Layang-Layang, Sayong and Linggiu River before converge to Johor River and flows southeast to Johor Straits with the total length ~122.7 km long. Among other tributaries of Johor River consist of Pengeli River, Semangar River, Lebak River, Telor River, Panti River, Tembioh River, Permandi River, Seluyut River, Berangan River, Tiram River and Lebam River (see Figure 3.3).

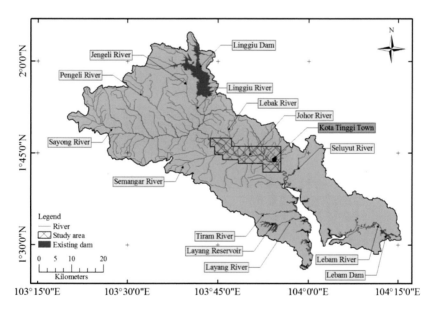

Figure 3.3: River system in Johor River Basin

Johor River and its tributaries are also important source of providing clean water not only for the Johor but also for the Singapore. To ensure adequate water supply, especially during drought season, there are three dams which were built in this basin namely as Linggiu dam, Lebam dam and Layang reservoir. The largest reservoir between these dams is the Linggiu dam which was located at the upstream of Johor River which is managed by Public Utility Board of Singapore. The dam was created across a tributary of the Johor River to discharge water during dry weather to regulate the flow in Johor River and prevent seawater from flowing in at the water treatment facilities. While, Lebam dam and Layang reservoir located at the downstream of Johor River.

In general, the river basin is flat except at the northern and eastern region of this basin. Based on the DEM from shuttle radar topography mission (SRTM), the highest elevation in the basin is 985 m and the lowest is -6 m (see Figure 3.4). The downstream part is covered with wetlands and swampy area. Generally, Johor River is a meandering channel with most of the areas is prone to have neck or chute cutoff with high sedimentation (Shafie 2009).

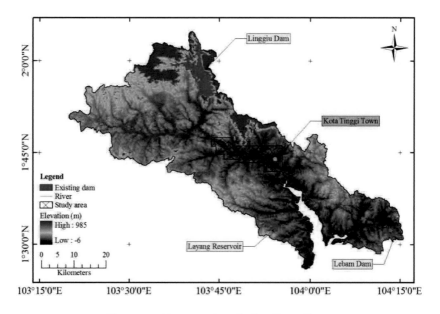

Figure 3.4: Topography of Johor River Basin

3.1.3 Climate

Johor River Basin receives equatorial climate with constant high temperatures and high relative humidity. The temperatures vary little throughout the year with an annual average temperature of 27°C and a mean relative humidity of 82%. As shown in Table 3.1, the rainfall pattern in Johor River Basin is influenced by two monsoon regimes, Northeast Monsoon and Southwest Monsoon.

Table 3.1: Monsoon regimes in Johor River Basin, Malaysia

Monsoon	Period	Characteristics
Northeast	November – March	Winds 2-20 knots up to 30 knots during cold surges period affecting east coast area. Heavy rainfall.
Southwest	May – September	Winds below 15 knots affecting west coast area. Drier weather.

The rainfall is usually greatest in the months of November to January but rain falls in all months of the year. Rainstorms are short, intense and generally have a limited spatial extent. The rainfall distribution over the catchment of Johor River is 3,000 mm of the north and north-west, and about 2,000 mm in the coastal region with an average of about 2,500 mm.

3.1.4 Land use

The land use conditions within the basin as a whole is covered by the natural forest, rubber and oil palm plantations. There are also areas of mining activities within the basin mostly sand mining. Base on the land use up to the year 2006 obtain from Department of Town and Country Planning, Malaysia (DTCP), about 18.56% of the area within Sungai Johor basin is covered by forest and 57.28% was in agriculture.

Most of the agriculture area is planted with oil palm and rubber tree. Other land uses are for built up area, swamp, water body and others (see Figure 3.5).

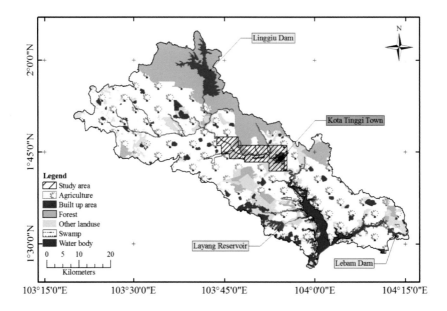

Figure 3.5: Present land use within Johor River Basin

As shown in Figure 3.6 by year 2020, the forest area will be reduced about 26.83% from the total area of the basin. For the built up area, the total area in year 2003 was 94.19 km², and it projected by the year 2020 the total area will be increased to 473.49 km².

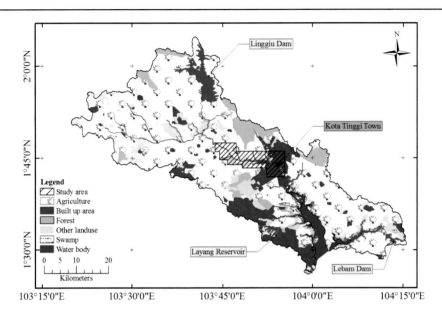

Figure 3.6: Future land use within Johor River Basin

3.1.5 Flood issues

The Johor River Basin has been subjected to repetitive flooding during its history. Most of the severe floods occurred during the north-east monsoon period, which brought large volume of runoff to the relatively large basin of Johor River. During December 2006 and January 2007 floods, it has been reported that Typhoon Utor has been associated to the series of floods that hit Malaysia, Singapore and Indonesia.

Figure 3.7: Condition at Kota Tinggi town, a) normal condition; and b) 2006/2007 flood event

37

According to DID (2009), the first flood event occurred from 19th to 24th December 2006 which widespread rainfall of about 350 mm was recorded. The second flood event happened from 11th to 17th January 2007 with a wide spread of heavy rainfall of 400 mm. The flood inundated Kota Tinggi town with a depth of about 3 m. While, 14,864 flood victims from 3,303 families and 15,660 victims from 3,483 families were evacuated in Kota Tinggi during first flood event (19-24 December 2006) and second flood event (11-17 January 2007), respectively (source: www.jkmnj.gov.my).

Figure 3.8: Historical flood event in Kota Tinggi town, a) normal condition; b) flood in December 1969; c) flood in December 2006; and d) flood in January 2007

Besides the recent two major floods, Kota Tinggi town also experienced occurrences of major historical floods in December 1948, December 1969, January 1970, November 1979, December 1982, November 1989, December 1983, December 1991, December 2003, December 2004 and March 2004 (DID 2009).

3.2 Data Availability

3.2.1 Hydrological data

There are about 23 numbers of rainfall stations, 5 numbers of water level station and 1 numbers of stream flow station within the Johor River Basin. All the hydrological stations were maintained by Department of Irrigation and Drainage, Malaysia (DID) (see Figure 3.9).

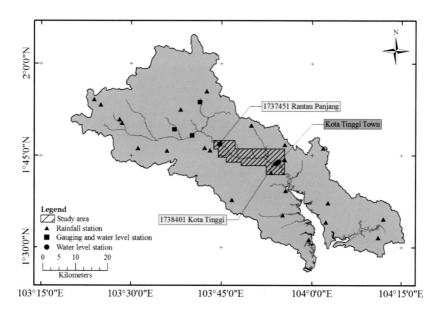

Figure 3.9: Hydrological station within Johor River Basin

However, for this study only 2 numbers of water level station and 1 number of stream flow station were consider for the purpose of calibration and verification of the hydraulic model. Table 3.2 and the locations of these stations are shown in Figure 3.9.

Table 3.2: Hydrological station use in this study

ID Number	Station Name	Type of Station
1738401	Sungai Johor at Kota Tinggi	Water level
1737451	Sungai Johor at Rantau Panjang	Water level and stream flow

The models were used to simulate two hydrological events: i) December 2006 flood (characterised by ~375 m^3s^{-1} peak flow) and ii) January 2007 flood (characterised by ~595 m^3s^{-1} peak flow). Both discharge data were measured and recorded at the Rantau Panjang hydrological station. The hydrological data required for this study were collected from the Hydrology and Water Resources Division of DID, Malaysia.

3.2.2 Topography data

For this study area, five type of topography data which are available either commercial or non-commercial (freely available). The following will describe the type of topography data used in this study.

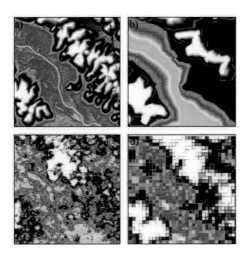

Figure 3.10: Type of DEMs used in this study based on: a) LiDAR; b) Topography map; c) ASTER data; and d) SRTM data

a.   Light Detection and Ranging (LiDAR)

The high resolution digital elevation model (DEM) of the entire case study was built by the DID in 2008 using the Light Detection and Ranging (LiDAR) survey method with a horizontal resolution of 1 m. This survey method is using a remote sensing technology that is used to measures x, y and z coordinates of terrain. These works were carried out using HARRIER 56/G3 mounted onboard a light aircraft at average flying height of 7000 m and swath width of approximately 600 m. It was reported that the vertical accuracy of this topographical data is 0.15 m (see DID, 2009). However, to produce high-resolution DEM data (such as LiDAR) of flood modelling will be required significant time and higher cost.

b.   The Shuttle Radar Topography Mission (SRTM)

The SRTM is a collaboration programme between National Aeronautics and Space administration (NASA) and National Geospatial-Intelligence Agency (NGA) of United States of America. The topographic data provided by these agencies are covered globally for all area of earth surface from 56ºS to 60ºN. There are two types of resolution of data products available: (i) 30 m resolution, and (ii) 90 m resolution. Nevertheless, for the 30 m resolution it's only available for the United States territory, whereas for the remaining of the world is the latter. Currently, the latest version of the SRTM DEM known as SRTM Version 4 can be downloaded freely from www.cgiar-sci.org. It was reported that the SRTM finished data meet the absolute horizontal and vertical accuracies of 20 m and 16 m, respectively. Noticeably, evaluation study carry out by Sander (2007) find out that SRTM DEM are better at flat floodplains compare than in high terrain.

c.   Advanced Spaceborne Thermal Emission and Reflection Radiometer (ASTER)

The Advanced Spaceborne Thermal Emission and Reflection Radiometer (ASTER) is a combined effort between NASA, Japan's Ministry of Economy, Trade and Industry (METI), and Japan Space Systems (J-spacesystems). As highlighted by Cuartero et al (2004), this type of DEM which generated by automatic correlation of ASTER stereo image data obtained an error in a range of 10 to 30 m. Furthermore, ASTER DEMs are known be very accurate in near flat and smooth slope topography. However, where areas are covered by forest, snow and limited sunlight exposure, the error a large (Eckert *et al.*, 2005; Tarekegn *et al.*, 2010). The ASTER DEMs were on the public domain and can be obtained from http://earthexplorer.usgs.gov/

d.   Cartography map

In Malaysia (particularly in the study area), the topographic maps are available in different scales. For this specific study area, topographic maps of 1:25 000 are used for research purposes which have a spatial resolution of 20 m and contours intervals at 20 m. The elevation data from this type of topographic maps are based on ground survey. However, due to the low coverage of data in such maps, the use of DEM produced may leads to the higher inaccuracies in prediction of flood extent.

e.   River cross section survey

The survey was carried out through a ground survey method across the river at average width range from 70 to 300 m with the spacing between cross section at approximately 1000 m. The purpose of this survey is to enable sufficient coverage data for the area under water which may not be captured by others survey method such as radar wave and laser altimetry.

There are also three numbers of bridges are identified located along this study reach. Information available for these bridges includes (i) river cross section surveys at the downstream and upstream of the bridges, and (ii) detailed bridge geometry and pier position.

Chapter 4
1-D hydraulic modelling: the role of cross-sections spacing

Large parts of this chapter has been published as:

Md Ali, A., Di Baldassarre, G., and Solomatine, D. P. (2015). Testing different cross-section spacing in 1-D hydraulic modelling: A case study on Johor River, Malaysia. Hydrological Sciences Journal. 60 (2), 351-360

4.1 Introduction

According to the Emergency Events Database (EM-DAT), more than 3,000 flood disasters were reported all around the world between 1992 and 2011. Those flood events were estimated to affect more than 2 billion people, and cause some 150,000 fatalities and economic losses around 480 billion USD. Depending on the societal context (Berz 2000), flood disasters may lead to damages of infrastructures and losses of human lives (Di Baldassarre *et al.*, 2010). A possible way to reduce the impact of flood disasters is the mapping of flood prone areas to raise risk awareness and support sustainable land-use planning and urban development (Horritt *et al.*, 2007). Flood mapping is also a useful source of information in rescue and relief missions. In 2007, the European Union (EU) has adopted the Directive 2007/60/EC (known as the Flood Directive) with the main objective to reduce and manage flood risk. To this end, all the Member States need to prepare flood hazard and risk maps by the end of 2013 (van Alphen *et al.*, 2009).

Despite the development of a plethora of two-dimensional (2-D) hydraulic models (e.g. Franques and Yannitell, 1974; Cunge 1975; Horrit and Bates, 2002; Horrit *et al.*, 2006; Tayefi *et al.*, 2007; Cook and Merwade, 2009; Castellarin *et al.*, 2011), one-dimensional (1-D) hydraulic models remain widely used for flood risk studies (Baptist *et al.*, 2004; Hooijer *et al.*, 2004; Di Baldassarre *et al.*, 2009). A number of studies have showed that the performance of 1-D model are often very close to the one of a 2-D model provided that the topography of the river and floodplain is properly represented (e.g. Horrit and Bates, 2002; Castellarin *et al.*, 2009; Cook and Merwade, 2009).

In terms of topographical data, 1-D modelling requires a certain number of cross-sections to represent the river channel and its surrounding topography (Cook and Merwade, 2009). Nevertheless, only a few guidelines are available to assist the modeller in determining the location or spacing between the cross-sections (Samuels 1990; 1995; Castellarin *et al.*, 2009; Fewtrell *et al.*, 2011). Common sense suggests that locating a river cross-section at every river bend and at upstream/downstream of every structure across the river may produce better result.

However, having cross-section data at finer spacing or at every river meander (i.e. river bend) and at upstream/downstream of structures (e.g. bridges) will increase the cost of the topographical surveys. Samuels (1990) recommended certain guidelines for the selection of cross-section locations for 1-D hydraulic models. These guidelines were based on a combination of common sense, practical experience and mathematical equations. Samuels' guidelines aimed to provide a correct reproduction of the backwater effects and an accurate representation of the physical waves. In general, Samuels (1990) emphasised to include the following locations of river cross-sections:

(i)   At both ends of the river reach (upstream and downstream of the reach);

(ii)   At the upstream/downstream of the river structures (i.e. bridge or weir);

(iii) At all discharge and water level stations along the reach; and

(iv)   At sites which are important to the modeller.

In addition to these "essential" cross-sections, Samuels (1990) recommended the following equations in determining the spacing of the cross-sections:

$$\Delta x \approx kB \tag{4.1}$$

where Δx is the spacing between two cross-section spacing; B is bankfull surface width of the main channel and k is constant (with the recommended range between 10 to 20).

For an appropriate estimation of backwater effects in subcritical flow, Samuels (1990) suggested:

$$\Delta x < 0.2 \frac{(1 - Fr^2)D}{s} \quad \text{or}$$

$$\Delta x < 0.2 \frac{D}{s} \qquad \text{if Froude number, } Fr^2 = 0 \tag{4.2}$$

where D = bankfull depth of flow, and s = main channel slope. Over this length, the backwater upstream of a control (as well other disturbance) decays to less than 0.1 of the original value.

For unsteady flow condition, an additional equation was suggested where two type of waves might be considered: the flood wave and the tide wave propagating along an estuary. Where c = propagation speed of wave; T = period of wave and N_{gp} = number of grid points (varies from 30 to 50)

$$\Delta x < \frac{cT}{N_{gp}} \qquad (4.3)$$

Finally, the minimum spacing between the cross-sections can be determined due the influence of the rounding error as follows:

$$\Delta x > \frac{10^{d-q}}{s\varepsilon_s} \qquad (4.4)$$

where q represents the number of decimal digits precision, d is the digits lost due to cancellation of the leading digits of the stage values; s is average surface slope and ε_s is relative error on surface slope. However, it was noted that, these guidelines were based on the condition of the rivers in United Kingdom.

To test the optimal cross-sectional spacing in 1-D hydrodynamic model, Castellarin et al. (2009) presented two case studies in United Kingdom and Italy. The main objective of their study was to evaluate the efficiency of the 1-D model by adopting some of the guidelines and equations recommended by Samuels (1990). As a result, for both selected river reach, the inaccuracy of the model in terms of mean absolute error (*MAE*) was found to increase as the spacing between the cross-sections increases. However, differences between the models were found to be relatively small (less than 0.10 m) and within the accuracy of computational hydraulic models (Samuels 1995; Di Baldassarre and Claps, 2011). Castellarin et al. (2009) concluded that while the minimum spacing of cross-section adopted through the Equations 4.2 and Equations 4.3 were appropriate for describing the hydraulic behaviour of the river reaches being studied, it was found inadequate when the hydraulic models is to be used for designing flood mitigation structures. The main limitation of the study by Castellarin et al. (2009) was that numerical results were not compared against observations. Instead, the results from the hydraulic model with the highest number of cross-sections were used as a benchmark/reference.

Cook and Merwade (2009) developed 1-D hydraulic modelling for two selected river reaches in USA and highlighted that the simulated flood extent (i.e. inundated area) becomes larger as the number of cross-section increases. Nevertheless, the effects are localized by omitting the structural details (as too much may give a misrepresentation of flooding around the structure). Such action will not affect the overall inundation extent over a reach. In the experiment, the authors also stressed that the increase or decrease of the inundation area are not only influenced by the series of the cross-section used, but also by the width of the cross-sections itself.

More recently, Akbari and Barati (2012) investigated the effects of varying time and space steps on the model performance in terms of Nash-Sutcliffe criterion through numerical tests. Their study showed that the effects of space steps on flood propagation results are less significant than the effects of time steps.

This chapter aims to explore the impact of cross-section spacing by using real world observations. In particular, the performance of different 1-D hydraulic models based on different cross-section spacing is evaluated (for the first time) using independent calibration and validation data of flood water level observations.

4.2 Methodology

4.2.1 Hydraulic Modelling

To perform hydraulic modelling, the Manning's n roughness coefficient for the river channel ranged from 0.020 to 0.080 $m^{-1/3}s$, while n for flood plain from 0.030 to 0.100 $m^{-1/3}s$ (Chow 1959). In order to test the sensitivity of the results to Manning's n roughness coefficient, the models were run using various n values within those ranges. In the absence of any knowledge of the prior distribution of the model parameters, a random distribution was assumed to select 5000 sets of Manning's n roughness coefficients value within these ranges for each hydraulic model. Apart from the Manning's n roughness coefficients and cross-section configurations, all other sources of data including the boundary conditions were unchanged for all simulations.

The models were used to simulate two hydrological events: the December 2006 flood event with a peak flow ~ 375 m^3s^{-1} was used in the calibration stage. Whereas, the January 2007 flood event, with a peak flow ~ 595 m^3s^{-1} was used for validation of the model. Both discharge data were measured and recorded at Rantau Panjang hydrological station.

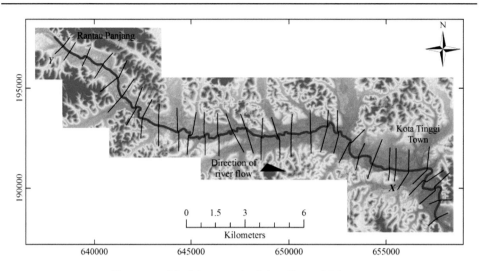

Figure 4.1: Model set-up for Johor River, Malaysia

The performance of each model were then was assessed by mean of *MAE*, where the simulated flood levels were compared with the observed during the December 2006 and January 2007 flood events at the reference cross-section X and cross-section Y (see Figure 4.1).

4.2.2 Cross-section spacing

In this chapter, five hydraulic models of the Johor River with different spacing of cross-sections were developed: Jhr XS1, Jhr XS2, Jhr XS3, Jhr XS4 and Jhr XS5. All models include the reference cross-section at point X and point Y as well as the upper and lower cross-section used for upstream and downstream boundary condition (see Figure 4.1). The cross-sections were derived by integrating the ground survey of the river channel and Light Detection and Ranging (LiDAR) digital elevation model (DEM) of the floodplain areas.

Table 4.1: Approximate of Johor River characteristics

Bankfull depth of flow, D (m)	Average main channel slope, $S_{avearge}$	Average Froude number, $Fr_{average}$	Period of wave, $T(s)$	Propagation speed of wave, c (m/s)
5.0 - 8.5	0.26-10-3	0.08-0.35	691200	0.36

Based on river characteristics shown in Table 4.1 and according to both Equations 4.2 and 4.3, the minimum spacing between the cross-sections for Jhr XS4 and Jhr XS5 is around 5,000 m and 8,000 m. More details of each model are presented in Table 4.2 and Figure 4.2.

Table 4.2: Details of geometric configuration for each model for the study reach

Model Code	Nos. of cross-section	Approximate spacing between cross-section	Description
Jhr XS1	41	~ 1000 m	32 numbers of cross-section spacing at ~1000 m based on modeller judgement.
			3 numbers of cross-section for each bridges crossing at 3 locations which include bridge detailing and upstream/downstream of the bridge.
Jhr XS2	32	~ 1000 m	Modeller own judgement
Jhr XS3	16	~ 2000 m	Modeller own judgement
Jhr XS4	8	~ 5000 m	Based on Equation 4.2
Jhr XS5	6	~ 8000 m	Based on Equation 4.3

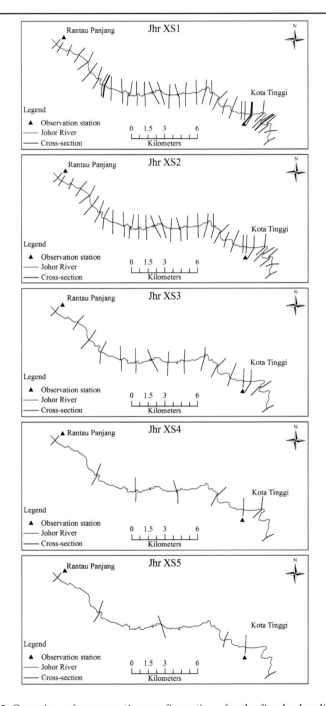

Figure 4.2: Overview of cross-section configurations for the five hydraulic models

4.3 Results and Discussion

This section is divided into three parts. The first part describes the performance of the 1-D hydraulic model for each geometric configurations. The second part discusses the influence of geometric configuration to flood inundation in terms of maximum water surface elevation, area of flood extent and flood depth. The third part highlighted the differences of the model output by inclusion/exclusion of the bridges structure in the hydraulic models.

4.3.1 Model performance

To investigate the influence of different cross-section spacing on the performance of 1-D model, five models were built as mention in section 4.2.2. Based on the simulation of the December 2006 flood (calibration event), Figure 4.3 shows the values of the *MAE* obtained with the five model configurations.

Then, the five best fit models (i.e. the five models using the calibrated Manning's *n* roughness coefficient) were used to simulate the January 2007 flood (validation event). Table 4.3 shows the average *MAE* of each model for the simulated water levels at the cross-section *X* and cross-section *Y*. It was found that the *MAE* value for different models does not change significantly as spacing between the cross-sections increased from 1,000 m to 8,000 m.

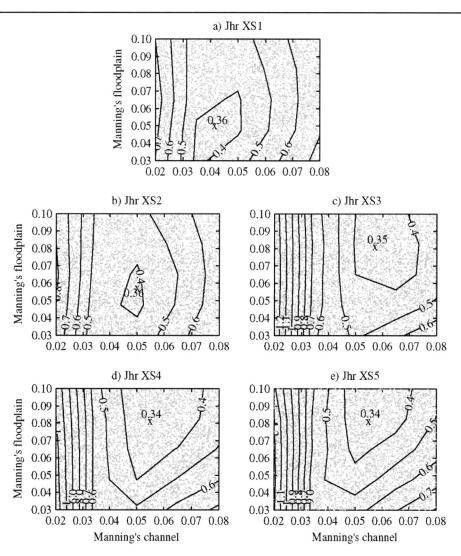

Figure 4.3: Calibration of the five hydraulic models: *MAE* versus the two model parameters for the December 2006 flood event. *x*-axis represent Manning's *n* channel and *y*-axis represent Manning's *n* floodplain

The results of this study as summarised in Table 4.3, confirm that the equations recommended by Samuels (1990) provide a proper indication about cross-section spacing. In fact, although the models derived from different spacing of cross-sections, the differences of *MAE* is between 0.02 m and 0.13 m.

Table 4.3: Model results and average *MAE* obtained at the observation points X and Y.

Model code	Calibrated Manning's n coefficients		MAE (m) (validation)
	channel	floodplain	
Jhr XS1	0.042	0.051	0.46
Jhr XS2	0.050	0.057	0.41
Jhr XS3	0.057	0.081	0.52
Jhr XS4	0.054	0.081	0.54
Jhr XS5	0.055	0.081	0.49

4.3.2 Comparing flood water profiles and inundation maps

The results of the hydraulic models can be used to produce flood water profiles (Brandimarte and Di Baldassarre, 2012) as well as flood inundation maps (Horritt and Bates, 2002) to support flood risk studies. The former are useful for the design of flood protection measures (e.g. dikes and levees), while the latter can allow the identification of flood prone areas.

Figure 4.4 shows the flood water profiles in terms of water surface elevations obtained with the five best fit models. The visual comparison of these five flood water profiles pointed out that, although small differences were found in terms of performance in simulating flood water levels at the cross-section X (see previous section 4.3.1), that significant differences arise in terms of the spatial distribution of the flood water profiles. This is caused by the absence of topographical information, and therefore geometric input and output, along the river. This outcome suggests that a detailed topographical survey is required when hydraulic modelling is meant to support the design of flood mitigation structures, such as levees (see also, Castellarin *et al.*, 2009).

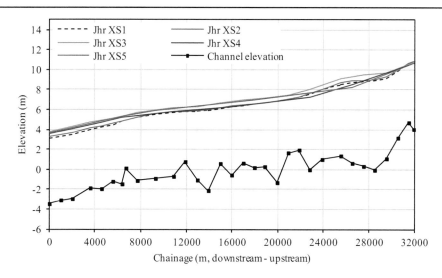

Figure 4.4: Flood profile: maximum water surface elevation along the Johor River for the five hydraulic models

Figure 4.5 shows the flood inundation maps obtained with the five hydraulic models using the calibrated Manning's n roughness coefficient, while Table 4.4 reports the corresponding inundated area. This comparison showed that the error in simulating water elevations at the points X are not significant as differences in MAE are within 0.04 m (see Table 4.4), which is less than the accuracy of LiDAR data (around 0.30 m). In contrast, the resulting flood inundation patterns are rather different (see Figure 4.5) as the flood extent mapping is strongly affected by the cross-sections used as geometric input of the hydraulic model.

Table 4.4: Impact of geometric configurations on 1-D hydraulic simulation.

Model Code	Approximate spacing between cross-section	*MAE* (m) at cross-section *X*	Total inundated area (km²)
Jhr XS1	~ 1000 m	0.36	25.64
Jhr XS2	~ 1000 m	0.38	24.58
Jhr XS3	~ 2000 m	0.35	23.47
Jhr XS4	~ 5000 m	0.34	22.14
Jhr XS5	~ 8000 m	0.34	18.26

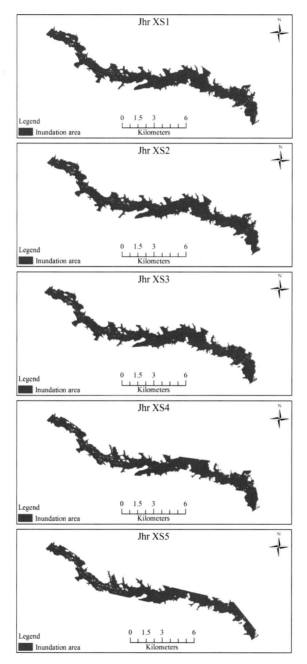

Figure 4.5: Effect of different cross-section configurations in flood inundation patterns

4.3.3 Representation of bridge structures in the model

The following section investigates the influence of geometric representation of the bridges in this study area. Model Jhr XS1 represent the model with a bridges, while Jhr XS2 represent the model without the bridges. More details of these two models are as mentioned in Section 4.2.2. The Jhr XS1 calibrates to *MAE* of 0.36 m with floodplain Manning's *n* roughness of 0.051 and a channel Manning's *n* roughness of 0.042, whereas the Jhr XS2 calibrates to *MAE* of 0.38 m with floodplain Manning's *n* roughness of 0.057 and channel Manning's *n* roughness of 0.050.

Further comparison was also made to assess the effects of the cross-sections configuration by means of inclusion/exclusion of the cross sections or hydraulic structures details on flood hazard map. Figure 4.6 and Figure 4.7 represent the calculated flood hazard area for each simulation case.

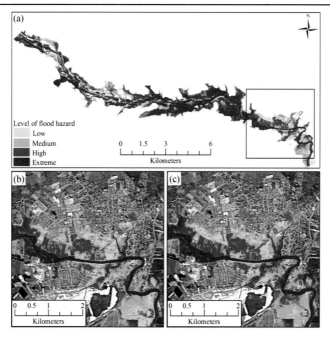

Figure 4.6: (a) Flood hazard map for the study area; (b) and (c) an enlarged area [indicated by black box in (a)] to illustrate the differences of flood hazard map by inclusion/exclusion of bridge structures

Based on the simulations, the largest flooded area is that resulting from the Jhr XS1 model, indicating that some area would be more prone to inundation of flood at this magnitude, whereas Jhr XS2 model predicted that this area are less inundated. The comparison between each hazard category shows that the most differences occurred in the low and extreme hazard categories. Jhr XS1 gives a lower percentage of low hazard area at 8.51% compared to 10.24% for Jhr XS2. Under an extreme hazard category, Jhr XS1 gives the highest percentage at 46.81% compared to 41.99% for Jhr XS2.

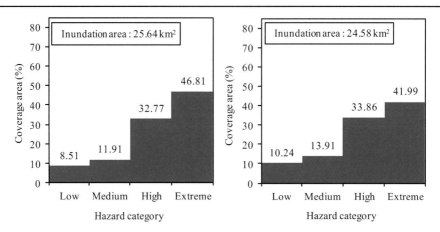

Figure 4.7: Histogram of coverage area for each flood hazard category: left panel represent Jhr XS1; and right panel represent Jhr XS2

The differences between the estimation of the hazard area using different simulation case, highlights the uncertainty in determining an accurate estimating of the flood hazard area. This can be due to the way that the flooded area is determined by interpolation between the cross section data. Therefore, there is missing important data in between cross sections during a changes or reduction in the number of cross sections which can introduce a certain error in the representations of floods on floodplain. Here, the exclusion of the bridge structural details in the geometric description, represented by Jhr XS2 model do have effects on the flood hazard categorization.

As such, the responses of the model with a bridge and without a bridge suggesting that although the bridge may have a significant impact on the water surface elevations or backwater effect, but it's does not reflect to the results of the calibration and validation. In fact this indications show that this study reach and selected flood event can be analyzed without having a bridge detailing. However, a part of contribution to this similarity may due to the constraints of the floodplain which determines the water levels.

In particular, in HEC-GeoRAS the flood inundation extent is generated through a semi-automated process, whereby the topography is subtracted from the water surface elevations. The domain of these inundation maps is created by linearly joining the water surface extent points on each cross-section as depicted in Figure 4.8. Therefore, the resulting inundation extent is related to the placement and the width of the cross-sections as well as definition of the geometry of the river and floodplain, which depends on the detail of the topographic information, i.e. number of cross-sections.

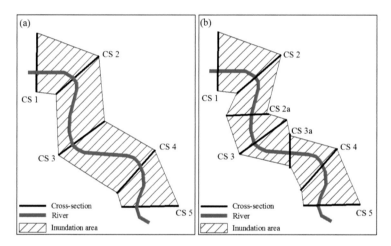

Figure 4.8: Effect of cross section locations on domain of flood inundation in HEC-GeoRAS

In summary, this additional analysis shows that if the model is meant to simulate inundation patterns of flood profiles, more cross-sections are needed to capture the spatial heterogeneity of the model domain and the consequent heterogeneity of the model results.

4.4   Concluding remarks

This chapter investigated the impact of cross-section spacing on the accuracy of 1-D hydraulic modelling of floods by testing the few literature guidelines via independent calibration and validation exercises.

The performance of the five models (characterised by different cross-section spacing ranging from around 1000 m to around 8000 m; see Table 4.1) did not show significant differences. For instance, the differences in model performance were between 0.02 m and 0.13 m for the validation event of January 2007. This confirms the findings by Castellarin et al. (2009) as the models developed according to Samuels (1990) performed as good as the models with a higher number of cross-sections.

Although model performance at the water level station X and Y is not significantly affected by model spacing, significant different results were obtained in terms of flood extent maps and flood water profiles. This indicates that detailed topographical survey is still needed for the design of flood mitigation structures (e.g. levees) and the identification of flood prone areas.

It is also found that the effect of exclusion of the bridge structures and reduction in the number of river cross sections in a 1-D model may not give a significant impact on the overall inundation extent and water surface elevation, but this exclusion however, does leads to a misrepresentation of flood hazard categories which are important in the flood risk management. However, those differences can also be partly attributed from the domain definition of the hydraulic models when performing the hydraulic simulations.

Chapter 5
2-D hydraulic modelling: the role of digital elevation models

5.1 Introduction

The prediction of inundation area and knowledge flood dynamics is a useful tool in floodplain and flood risk management (Horritt *et al.*, 2007). In the past, to use a one dimensional (1-D) hydraulic model is a common practice among hydraulic modeller compared to the two dimensional (2-D) hydraulic models. This is due to the fact that 2-D models have been confined by the inadequacy of high quality topographic data (Horrit and Bates, 2001; Mark and Bates, 2000; Yu and Lane, 2006; Fewtrell *et al.*, 2008). However, with development of topographic data collection techniques to form digital elevation models (DEMs), from conventional (e.g. ground survey) to airborne remote sensing techniques (e.g. synthetic aperture radar [SAR]; light detection and ranging [LiDAR]), and with availability of computational resources, the later hydraulic model, may produce more accurate simulated results.

Flood inundation map is always represented as a deterministic map. Without a careful understanding of the uncertainty variables such as DEMs, the accuracy of the maps produced may be affected. Uncertainty of DEMs in flood inundation mapping amongst others is generated by the difference in the sources of DEMs used, the spatial resolution of the DEMs (cell size) and technique used for interpolation of the DEMs data. The consequence of uncertainty of DEMs into flood inundation mapping has been emphasis by Merwade et al. (2008) where it will affects the estimated discharge value from hydrological analysis, affects the water surface profile from hydraulic models, and the extent of flood boundaries in x-axis. For instance, Mark et al. (2004) recommended the use of high resolution of DEMs as input data for 1-D urban hydraulic modelling ranges between 1 to 5 m. The author highlighted the used of finer resolution DEM (i.e. 1 m) is necessary for a detailed analyses and does provide a much better visual presentation of the flood extent. In contrast, the used a coarser DEM (i.e. 5 m) will allow quick assessment of the model.

Contrary to the fact that availability of high resolution DEMs such as LiDAR have been enhanced in the last decade, a lower resolution than the spatial resolution (higher resolution) more often been used especially when dealing with a wider study area. Although this process will contribute to uncertainty in flood inundation mapping, it allow the model calibration and sensitivity analysis to be completed within a reasonable time frame. However, the re-scaling of LiDAR DEMs in hydraulic flood modelling give possibility of loss of important data or features (i.e. dykes) which may affect the model results. Furthermore, re-scaling process of raster data via Geographic Information System (GIS) may also deteriorate the accuracy of the DEMs.

To date, ample researches related to DEMs analysis in hydraulic modelling of flood (either 1-D or 2-D) have considered different resolutions and re-scaling technique of DEMs in model outputs such as in the production of flood inundation and flood profiles. Bates and De Roo (2000) demonstrated the ability of LISFLOOD-FP to predict flood inundation extent by using different resolution of DEMs as the input data. This model was specifically developed to predict flood inundation while minimising or ignoring the representation of the process. By using ARC-INFO GIS software, the original DEMs (with 5 m resolution) which is in triangulated irregular network (TIN) based format was converted into raster coverage at 25, 50 and 100 m resolutions. The models were set up in steady and dynamic state and used 1995 flood event inflow hydrograph as the boundary condition. While, aerial photo and SAR flood inundation imagery for the particular flood event was used for validation. Part of the results from this study indicated that the model set-up using 25 m resolution performed better than the other two coarser DEMs (i.e. 50 and 100 m).

Horrit and Bates (2001) demonstrated the effects of the spatial resolution on a raster based model. The DEM is provided by LiDAR survey data at a resolution of 10 m. In this study, local averaging technique was adopted to re-scale the original DEMs to 20, 50, 100, 250, 500 and 1000 m resolution. The authors concluded that the model reached a maximum performance at resolution of 100 m, while the model with 500 m resolution proved to be adequate for the prediction of water level.

Yu and Lane (2006) have investigated the impact of the choice of re-scaling strategy on the output of the LISFLOOD-FP model results. Using the standard re-scaling strategies (i.e. nearest neighbour interpolation, bilinear interpolation, mean and cubic spline convolution), the DEMs used in the study were re-scaled from 2 m (original resolution) to 4, 8 and 16 m. The authors concluded that there was not much difference of the model outputs associated to the re-scaling strategy and found inconsistent results of model simulations across the re-scaled DEMs used. As addition, the authors highlighted that to obtain effective predictions of flood inundation extent, the high resolution DEMs need to be coupled with more advance inundation process.

In Glasgow, UK, Fewtrell et al. (2008) set up a 2-D model, LISFLOOD-FP, using different resolutions of LiDAR DEMs. By adopting the same standard re-scaling strategy used by Yu and Lane (2006), the original LiDAR DEMs were re-scaled from its original resolution, 2 m to 4, 8 and 16 m resolution. From the results, the author concluded that the bilinear interpolation techniques appear to provide the most accurate results. The authors also suggested that for a flood modelling in urban environments, a coarse resolution model DEMs is limited by the representation of detailing in coarse model grids. Furthermore, the results of model simulations in this study are consistent across model scales and contrary to previous findings conducted by Yu and Lane (2006).

In particular, the result presented in this chapter aims to fulfil two objectives. The first objective is to evaluate the effect of different re-scaling techniques on model performance and inundation area. Secondly, to investigate the influence of different resolution on the model performance and inundation area which will be evaluated relative to benchmarking high resolution DEMs.

5.2   Differentiation of DEMs re-sampling technique

In this chapter, the aggregate functions are applied using Spatial Analysis Tools in ArcGIS software. Aggregate functions aggregated a series of cells to the same value to produce a single, coarser resolution cell. Based on a specified aggregation technique (i.e. Sum, Min, Max, Mean, or Median) in aggregate function, the DEMs re-sampled the input raster to coarser resolutions (i.e. output raster). As shown in Figure 5.1, function of Min has been specified as aggregation technique. Thus, the minimum value within the specified input cell factor, or resolution, of four [see Figure 5.1a], of four is determined for each cell in the output raster [see Figure 5.1b].

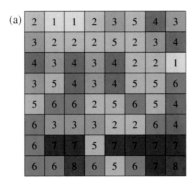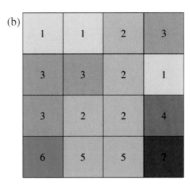

Figure 5.1: Illustration of aggregation techniques as a function of Min (a) input raster, 1 x 1 resolution; (b) output raster, 4 x 4 resolution

In order to test the significance of scale, the original LiDAR DEM with resolution of 1 m was re-sampled into three coarser resolution raster DEMs at 30, 90 and 120 m. As shown in Table 5.1, for each re-sampled DEMs namely as Raster Type A and Raster Type B, two different re-sampling techniques were applied.

Table 5.1: Re-sampling technique for the DEMs

DEM type	Aggregation technique	
	Channel	Floodplain
Raster type A	Min	Max
Raster type B	Mean	Mean

For Raster Type A, the channel DEM was specified as a function of Min, while the floodplain DEM was specified as a function of Max. Whilst for Raster Type B, both channel and floodplain DEMs were specified as function of Mean.

In total, six models were developed according to the six sets of DEMs with different resolution and different re-sampling technique (see Table 5.2).

Table 5.2: Type of DEMs with different resolution

	Model code	
DEM resolution (m)	Raster type A	Raster type B
30	Jhr A30	Jhr B30
90	Jhr A90	Jhr B90
120	Jhr A120	Jhr B120

Assuming the high resolution simulation represent a set of benchmark predictions (here the DEMs with 30 m resolution were used as a benchmark model), it is possible to verify if coarser resolutions model results are consistent with the benchmark simulation results. Subsequently, model verification is undertaken by assessing the coarser model predictions of model performance and flood extent with respect to the benchmark model results.

In order to assess the sensitivity of the different models to the model parameters, the Manning's n roughness coefficients for all the models were randomly distributed from 0.02 to 0.08 $m^{-1/3}s$ for the river channel, and between 0.03 and 0.10 $m^{-1/3}s$ for the floodplain. 500 simulations were carried out for each model. The performance of the hydraulic models in producing the observed water levels was assessed by means of the Mean Absolute Error (MAE). The data from two recent major flood events that occurred along the Johor River in 2006 and 2007 were used for independent calibration and validation of the models. The model results are then compared in terms of the model performance and area of inundation extent.

5.3 Results

5.3.1 Model Calibration and validation

The results of the calibration exercise in simulating 2006 flood event are shown in Figure 5.2, for the two types of raster DEMs, at 3 different resolutions, using the measure of fit *MAE*. For the model based Raster Type A (see Figure 5.2a, 5.2b and 5.2c), the model generates a well defined peak in the response surface, with most models showed to be more sensitivity to the changing of Manning's *n* roughness coefficient of main channel than the Manning's *n* roughness coefficient of floodplain areas. In contrast, all models based on Raster Type B DEMs response shows more sensitive to floodplain friction compared to Raster Type A models (see Figure 5.2-D, 5.2e and 5.2f).

For example, at 90 m resolution, the optimum for Raster Type A (see Figure 5.2b) occurs at Manning's *n* value of 0.05 $m^{1/3}s$ for the channel, and 0.03 – 0.10 $m^{1/3}s$ for the floodplain respectively. Whereas for Raster Type B (see Figure 5.2e), best calibration is at values between 0.055-0.08 $m^{1/3}s$ (or lower) for the channel and 0.03-0.09 $m^{1/3}s$ or lower for the floodplain, as the optimum at this flow lies outside the parameter space. Furthermore, the results of the calibration showed that the best fit models based on Raster Type A DEM with different resolution (Jhr A30, Jhr A90 and Jhr A120) generally give a good performance in the *MAE* value from 0.19 to 0.34 m (see Table 5.3). While, Table 5.4 shows that the best fit models based on Raster Type B DEM (Jhr B30, Jhr B90 and Jhr B120) gives *MAE* value from 0.48 to 0.77 m.

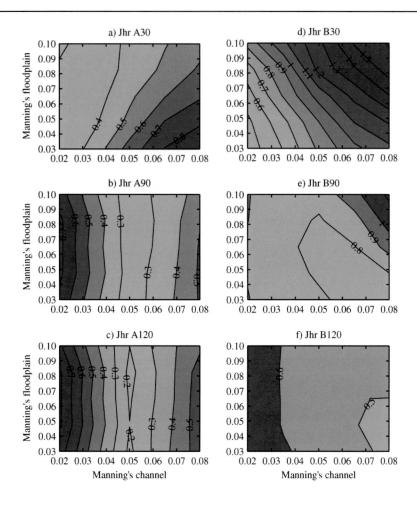

Figure 5.2: Contour maps of MAE from maximum simulated water level compared to the observed water level mapped over the friction parameter space using two types of raster DEM (Raster Type A and Type B) and 3 different DEM resolutions (30, 90 and 120 m)

In addition, this analysis showed that the accuracy of the models based on DEM Raster Type A is ascending as the resolution of DEMs decreased from 30 m to 120 m. Results from the 120 m model (see Jhr A120, Table 5.3) are surprisingly good, indicating that a low resolution model is capable of making reasonable estimation of water levels.

Table 5.3: Results of calibration and validation for DEM raster Type A

Model Code	Calibrated Manning's *n* roughness coefficient		MAE (m) (calibration)	MAE (m) (validation)
	Channel	Floodplain		
Jhr A30	0.034	0.068	0.34	0.68
Jhr A90	0.055	0.068	0.24	0.52
Jhr A120	0.050	0.076	0.19	0.65

Using the optimum Manning's *n* roughness coefficient from the best fit models, were then used to simulate the January 2007 flood event as a model validation. This was carried out for all models. Table 5.3 (represent models based on Raster Type A) and Table 5.4 (represent models based on Raster Type B) summarizes the *MAE* of each model. Although with a different resolution, it should be noted that the *MAE* for models based on Raster Type A remained at the value of *MAE* between 0.52 to 0.68 m. whereas for models based on Raster Type B, the *MAE* value shows an increase in model errors (*MAE* between 1.08 to 1.16 m).

Table 5.4: Results of calibration and validation for DEM raster Type B

Model Code	Calibrated Manning's *n* roughness coefficient		MAE (m) (calibration)	MAE (m) (validation)
	Channel	Floodplain		
Jhr B30	0.021	0.040	0.48	
Jhr B90	0.052	0.066	0.77	1.16
Jhr B120	0.064	0.072	0.51	1.08

The results from this analysis revealed that the reduction in the resolution of DEMs does not significantly affect the model performance. However, the used of different re-sampling methods as geometric input to the hydraulic model produces different model errors. For instance, at the resolution of 120 m, the model based Raster Type A, Jhr A120 (see Table 5.3) provided a *MAE* of 0.65 m, while the corresponding model based Raster Type B, Jhr B120 (see Table 5.4) provide a *MAE* equal to 1.08 m. Thus, the model based Raster Type A performance better than the model based Type B.

5.3.2 Flood simulation

Figure 5.3 shows floodplain boundaries delineated from DEMs of varying resolutions and re-sampling technique. The floodplain area of models based from Raster Type A (Figure 5.3a, 5.3b and 5.3c) tends to decrease with decreasing DEM resolution. Whereas, the models based from Raster Type B, shows insignificant variation in the floodplain area with respect to the decreasing DEM resolution (5.3d, 5.3e and 5.3f). From this analysis, the use of different re-sampling technique exhibits clear differences in the model output results. Inundation areas derived from Raster Type A models are fragmented with small patches, as compared to the relatively continuous inundation areas derived from the Raster Type B models. The differences are especially dramatic in the downstream area.

The reduced inundation area seen in Raster Type A models is due to the fact that higher resolution data enable sudden changes in elevation to be recorded more accurately compared to lower resolution data. A higher resolution DEMs would therefore detect and describe small sinks and peaks more precisely whereas lower resolution DEMs may even be unable to capture these small sinks or peaks. Nevertheless, differences in inundation areas due to the changes in resolution are relatively small, unlike the differences from using raster from different re-sampling technique.

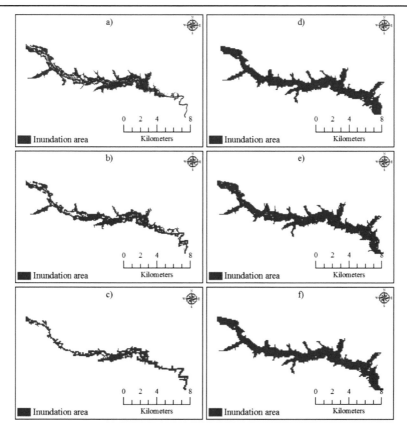

Figure 5.3: Flood inundation maps based on different resolution and re-sampling technique of DEMs: (a) Jhr A30; (b) Jhr A90; (c) Jhr A120; (d) Jhr B30; (e) Jhr B90 and (f) Jhr B120

In summary, this part of analysis suggests that although resolutions play some roles in the determining flood prediction results, it is not an influential factor, as inundation areas remained almost the same with similar sizes and shapes for DEMs at different resolution levels but from the same source (see model based Raster Type B, Figure 5.3d, 5.3e and 5.3f). By contrast, DEMs from different re-sampling technique produced vital discrepancies in simulation results. Therefore, how DEMs are re-sampled is critical in assessing flood impacts.

5.4 Conclusions

Having a fine resolution of DEMs are likely to create an accurate floodplains, however it often re-sampled to lower resolutions, or are used in analyses performed at lower resolutions by considering other attribute factors such as cost, computation ability, time and storing. Thus, re-sampling of the DEMs may become more common due to important of flood modelling at regional or global scale. In this chapter, effects of elevation data resolution and re-sampling technique on the model performance and flood simulation are analyzed across three different DEM resolutions. Two re-sampling techniques with three different aggregation functions (Min, Max and Min) are utilized for aggregating the LiDAR DEM from 1-m to 120-m.

The output of aggregating give an inconsistent result depending on the morphology: for the channel network, the Min functions are consistent; for the floodplain, function of Max is more suitable. Therefore, the selection of re-sampling technique is an important step that needs to be carried out in accordance with the intended use of the DEMs. It's also found out that lowering data resolution does not necessary lead to a poor performance of the model.

Chapter 6
1-D hydraulic modelling: the role of digital elevation models

This chapter has been published as:

Md Ali, A., Solomatine, D. P., and Di Baldassarre, G. (2015). Assessing the impact of different sources of topographic data on 1-D hydraulic modelling of floods. *Hydrology and Earth System Sciences*. 19, 1-13.

6.1 Introduction

In hydraulic modelling of floods, one of the most fundamental input data is the geometric description of the floodplains and river channels often provided in the form of digital elevation models (DEM). During the past decades, there has been a significant change in data collection for topographic mapping technique, from conventional ground survey to remote sensing techniques (i.e. radar wave and laser altimetry; e.g. Mark and Bates, 2000; Castellarin *et al.*, 2009). This shift has a number of advantages in terms of processing efficiency, cost effectiveness and accuracy (Bates 2012; Di Baldassarre and Uhlenbrook, 2012).

DEMs can be acquired from many sources of topographic information ranging from the high resolution and accurate, but costly, light detection and ranging (LiDAR) survey obtained from lower altitude to low-cost, and coarse resolution, space-borne data, such as advanced spaceborne thermal emission and reflection radiometer (ASTER) and shuttle radar topography mission (SRTM). DEMs can also be developed from traditional ground surveying (e.g. topographic contour maps) by interpolating a number of elevation points.

DEM horizontal resolution, vertical precision and accuracy varies considerably. These differences are attributed from different types of equipment and methods used in obtaining the topographic data. When used as an input to hydraulic modelling, the differences in the quality of each DEM subsequently result in differences in model output performance. In addition, re-sampling processes of raster data via Geographic Information System (GIS) may also deteriorate the accuracy of the DEMs. The usefulness of diverse topographic data in supporting hydraulic modelling of floods is subject to the availability of DEMs, economic factors and geographical conditions of survey area (Cobby and Mason, 1999; Casas *et al.*, 2006; Schumann *et al.*, 2008). To date, a number of studies have been carried out with the aim of evaluating the impact of accuracy and precision of the topographic data on the results of hydraulic models.

Werner (2001) investigated the effect of varying grid element size on flood extent estimation from a 1-D model approach based on a LIDAR DEM. The study found that the flood extent estimation increased as the resolution of the DEM becomes coarser.

Horrit and Bates (2001) demonstrated the effects of spatial resolution on a raster based flood model simulation. Simulation tests were performed at resolution sizes of 10, 20, 50, 100, 250, 500, and 1000 m and the predictions were compared with satellite observations of inundated area and ground measurements of floodwave travel times. They found that the model reached a maximum performance at resolution of 100 m when calibrated against the observed inundated area. The resolution of 500 m proved to be adequate for the prediction of water levels. They also highlighted that the predicted floodwave travel times are strongly dependent on the model resolution used.

Wilson and Atkinson (2005) set up a two-dimensional (2-D) model, LISFLOOD-FP, using three different DEMs [contour dataset, synthetic-aperture radar (SAR) dataset, and differential global positioning system (DGPS)] used to predict flood inundation for 1998 flood event in the United Kingdom. The results showed that the contour datasets resulted in a substantial difference in the timing and the extent of flood inundation when compared to the DGPS dataset. Although the SAR dataset also showed differences in the timing and the extent, it was not as massive as the contour dataset. Nevertheless, the authors also highlighted a potential problem with the use of satellite remotely sensed topographic data in flood hazard assessment over small areas.

Casas et al. (2006) investigated the effects of the topographic data sources and resolution on one-dimensional (1-D) hydraulic modelling of floods. They found out that the contour-based digital terrain model (DTM) was the least accurate in the determination of the water level and inundated area of the floodplain, however the global positioning system (GPS)-based DTM lead to a more realistic estimate of the water surface elevation and of the flooded area. The LiDAR-based model produced the most acceptable results in terms of water surface elevation and inundated flooded area compared to the reference data. The authors also pointed out that the different grid sizes used in LiDAR data has no significant effect on the determination of the water surface elevation. In addition, from an analysis of the time-cost ratio for each DEMs used, they concluded that the most cost effective technique for developing a DEM by means of an acceptable accuracy is from LiDAR, especially for large areas.

Schumann et al. (2008) demonstrated the effects of DEMs on deriving the water stage and inundation area. Three DEMs at three different resolutions from three sources (LiDAR, contour and SRTM DEM) were used for a study area in Luxembourg. By using the HEC RAS 1-D hydraulic model to simulate the flood propagation, the result shows that, the LiDAR DEM derived water stages by displaying the lowest *RMSE*, followed by the contour DEM and lastly the SRTM. Considering the performance of the SRTM (it was relatively good with *RMSE* of 1.07 m), they suggested that the SRTM DEM is a valuable source for initial vital flood information extraction in large, homogeneous floodplains.

For the large flood prone area, the availability of DEM from public domain (e.g. ASTER, SRTM) makes it easier to conduct a study. Patro et al. (2009) selected a study area in India and demonstrated the usefulness of using SRTM DEM to derive river cross section for the use in hydraulic modelling. They found that the calibration and validation results from the hydraulic model performed quite satisfactory in simulating the river flow. Furthermore, the model performed quite well in simulating the peak flow which is important in flood modelling. The study by Tarekegn *et al.* (2010) carried out on a study area in Ethiopia used a DEM which was generated from ASTER image. Integration between remote sensing and GIS technique were needed to construct the floodplain terrain and channel bathymetry. From the results obtained, they concluded that the ASTER DEM is able to simulate the observed flooding pattern and inundated area extends with reasonable accuracy. Nevertheless, they also highlighted the need of advanced GIS processing knowledge when developing a digital representation of the floodplain and channel terrain.

Schumann et al. (2010) demonstrates that near real-time coarse resolution radar imagery of a particular flood event on the River Po (Italy) combined with SRTM terrain height data leads to a water slope remarkably similar to that derived by combining the radar image with highly accurate airborne laser altimetry. Moreover, it showed that this spaceborne flood wave approximation compares well to a hydraulic model thus allowing the performance of the latter, calibrated on a previous event, to be assessed when applied to an event of different magnitude in near real time.

Paiva et al. (2011) demonstrated the use of SRTM DEM in a large-scale hydrologic model with a full one-dimensional hydrodynamic module to calculate flow propagation on a complex river network. The study was conducted on one of the major tributaries of the Amazon, the Purus River basin. They found that a model validation using discharge and water level data is capable of reproducing the main hydrological features of the Purus River basin. Furthermore, realistic floodplain inundation maps were derived from the results of the model. The authors concluded that it is possible to employ full hydrodynamic models within large-scale hydrological models even when using limited data for river geometry and floodplain characterization.

Moya Quiroga et al. (2013) used Monte Carlo simulation sampling SRTM DEM elevation, and found a considerable influence of the SRTM uncertainty on the inundation area (the HEC-RAS hydraulic model of the Timis-Bega basin in Romania was employed).

Most recently, a study by Yan et al. (2013) made a comparison between a hydraulic model based on LiDAR and SRTM DEM. Besides the DEM inaccuracy, they also introduced the uncertainty analysis by considering parameter and inflow uncertainty. The results of this study showed that the differences between the LiDAR-based model and the SRTM-based model are significant, but within the accuracy that is typically associated with large-scale flood studies.

This chapter continues the presented line of research and deals with the assessment of the effect of using different DEM data source and resolution in a 1-D hydraulic modelling of floods. The novelty of this study is that both quality (in terms of accuracy) and precision (resolution) of the DEM are considered and their impact on hydraulic model results is evaluated in terms of both water surface elevations, inundation area and flood hazard maps.

6.2   Available data

6.2.1 Hydraulic Modelling

In this study, hydraulic modelling of floods was performed by using the HEC-RAS modelling system. To simulate unsteady open channel flow, HEC-RAS solves the full 1-D Saint-Venant Equations. The observed flow hydrograph at an hourly time step was used as upstream boundary condition, while the friction slope was used as downstream boundary condition. The next section reports the different sources of topographic data used to define the geometric input.

6.2.2 Digital Elevation Model

The required input data for the HEC RAS include the geometry of the floodplain and the river, which is provided by a number of cross sections. We identified several sources of DEM data for our study area (details are given below) with different spatial resolution and accuracy:

    i.   DEMs derived from an original 1 m LiDAR dataset (obtained from DID).

ii.   20 m resolution DEM generated from the vectorial 1:25000 cartography map obtained from Department of Irrigation and Drainage, Malaysia (DID) with a permission of the Department of Survey and Mapping, Malaysia (DSMP).

iii.   30 m resolution DEM derived from the globally and freely available ASTER data retrieved from the United States Geological Survey (USGS, http://earthexplorer.usgs.gov)

iv.   90 m resolution DEM derived from the globally and freely available SRTM data retrieved from a Consortium for Spatial Information (CGIAR-SCI, www.cgiar-sci.org).

Table 6.1: Information about the eight DEMs used as topographical input

Model code	DEM type	Resolution (m)
Jhr Ref	LiDAR	1
Jhr L2	(re-scaled from LiDAR, Jhr Ref)	2
Jhr L20	(re-scaled from LiDAR, Jhr Ref)	20
Jhr L30	(re-scaled from LiDAR, Jhr Ref)	30
Jhr L90	(re-scaled from LiDAR, Jhr Ref)	90
Jhr T20	Contour map	20
Jhr A30	ASTER	30
Jhr S90	SRTM	90

To analyse the influence of spatial resolution and separate it out from the impact of different accuracy, four additional DEMs were obtained by rescaling the original LiDAR DEM (1 m resolution) to the spatial resolutions of the DEMs derived from contour map (20 m), ASTER (30 m) and SRTM (90 m). Hence, a total of eight DEMs were used to explore the impact of different topographic information on the hydraulic modelling of floods (see Table 6.1 and Figure 6.1).

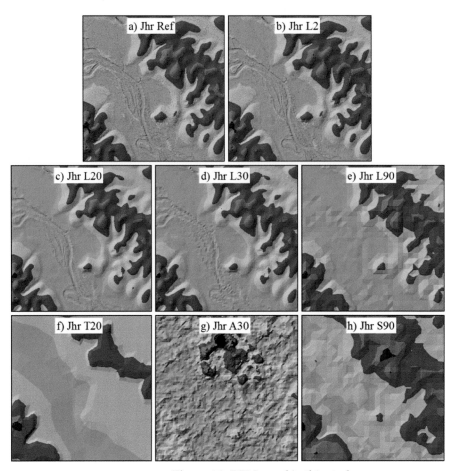

Figure 6.1: DEMs used in this study

6.3 Methodology

6.3.1 Evaluating the DEMs quality

At first, the vertical error of each DEM was evaluated through comparison between the topographic data and 164 GPS ground points taken at random positions within the study area. The value of each reference elevation points were extracted from the study area using GPS survey equipment. The reference elevation points were assumed to be the observed ground elevation. Based on common methods to quantify DEM error, several measures were calculated: Mean Error (*ME*) and Root Mean Square Error (*RMSE*):

$$ME_{DEM} = \frac{1}{n} \sum_{i=1}^{n} \left(Elev_{GPS_i} - Elev_{DEM_i} \right) \tag{6.1}$$

$$RMSE_{DEM} = \sqrt{\frac{\sum_{i=1}^{n} \left(Elev_{GPS_i} - Elev_{DEM_i} \right)^2}{n}} \tag{6.2}$$

where $Elev_{GPS_i}$ denotes as the value of reference elevation point extracted from GPS, $Elev_{DEM_i}$ is the value of elevation data of each DEMs, and n corresponds to the total numbers of reference data points used. The $RMSE_{DEM}$ give a measure of accuracy. It exhibit how far, on average; the observed values are from the assumed true value. While the ME_{DEM} show whether a set of measurements consistently underestimate (negative *ME*) or overestimate (positive *ME*) the reference value.

6.3.2 Model calibration and validation

Then, data from two recent major flood events that occurred along the Johor River in 2006 and 2007 were used for independent calibration and validation of the models. The estimated peak flow of the 2006 event is approximately 375 m³/s, while the one of the 2007 event is around 595 m³/s. Both discharge data were measured and recorded at Rantau Panjang hydrological station. The 2006 flood data were used for the calibration exercise, while the 2007 flood data were used for model validation.

To assess the sensitivity of the different models to the model parameters, the Manning's n roughness coefficients for all the models were sampled uniformly from 0.02 to 0.08 $m^{-1/3}s$ for the river channel, and between 0.03 and 0.10 $m^{-1/3}s$ for the floodplain, by steps of 0.0025 $m^{-1/3}s$. The performance of the hydraulic models in producing the observed water levels was assessed by means of the Mean Absolute Error (*MAE*).

6.3.3 Quantifying the effect of the topographic data source on the water surface elevation and inundation area (sensitivity analysis)

The effects of DEM source and spatial resolution were further investigated by examining the sensitivity of model results in terms of maximum water surface elevation (WSE), inundation area and floodplain boundaries. For this additional analysis, the model results obtained with the most accurate and precise DEM source (LiDAR at 1 m resolution) was used as a reference. For WSE analysis, each model was compared to the reference model (Jhr Ref, see Table 6.1) by means of the following measures:

$$MAE_{WSE} = \frac{1}{n}\sum_{i=1}^{n}\left|WSE_{Ref_i} - WSE_{DEM\,i}\right|$$ (6.3)

where WSE_{Ref_i} denotes the WSE simulated by the reference model, WSE_{DEM_i} the WSE estimated by the models based on each test DEMs (see Table 6.1), and x corresponds to the total number of cross-sections where model results were compared.

For comparison of the inundation area, the *F* statistic (e.g. Horrit *et al.*, 2007; Di Baldassarre *et al.*, 2009) quantitative indices were used. In particular, the inundation map obtained with the LiDAR based hydraulic model, Jhr Ref, was used as the reference map.

$$F = \frac{A}{A+B+C}$$ (6.4)

Where A refers to area that is both observed and predicted as inundated, B refers to area of inundation present in model but absent in observation, and C refers to area of inundation present in observation but absent in model. A value of 1 means a perfect match between observed and predicted areas of inundation, and a lower F indicates discrepancy between observed and predicted.

6.3.4 Uncertainty Estimation – GLUE analysis

In hydraulic modelling, multiple sources of uncertainty can emerge from several factors, such as model structure, topography, and friction coefficients (Aronica *et al.*, 2002; Trigg *et al.*, 2009; Brandimarte and Baldassarre, 2012; Dottori *et al.*, 2013). A methodological approach to estimate the uncertainty is the generalised likelihood uncertainty estimation (GLUE) methodology (Beven and Binley, 1992), a variant of Monte Carlo simulation. Although some aspects of this methodology are criticized in several papers (e.g. Hunter *et al.*, 2005; Mantovan and Todini, 2006; Montanari 2005; Stedinger *et al.*, 2008), it is still widely used in hydrological modelling because of its easiness in implementation and a common-sense approach to use only a set of the "best" models for uncertainty analysis (e.g. Hunter *et al.*, 2005; Shrestha *et al.*, 2009; Vázquez *et al.*, 2009; Krueger *et al.*, 2010; Jung and Merwade, 2012; Brandimarte and Woldeyes, 2013).

According to the GLUE framework (Beven and Binley, 1992), each simulation, i, is associated to the (generalized) likelihood weight, W_i, ranging from 0 to 1. The weight, W_i is expressed as a function of the measure fit, ε_i, of the behavioural models.

$$W_i = \frac{\varepsilon_{max} - \varepsilon_i}{\varepsilon_{max} - \varepsilon_{min}} \qquad (6.5)$$

where, ε_{max} and ε_{min} are the maximum and minimum value of *MAE* of behavioural models. To identify the behavioural of the models, a threshold value (rejection criteria) has been set as follows:

i.   simulations associated with *MAE* larger than 1.0 m; and

ii. Manning's *n* roughness coefficient of the floodplain smaller than the Manning's *n* roughness coefficient of the channel.

Then, the likelihood weights are the cumulative sum of 1 and the weighted 5th, 50th and 95th percentiles. The likelihood weights were calculated as follow:

$$L_i = \frac{W_i}{\sum\limits_{i=1}^{n} W_i} \tag{6.6}$$

For this study, the applications of uncertainty analysis considered only the parameter uncertainty and implemented for all DEMs based model.

6.4 Results and discussion

6.4.1 Quality of DEMs compared with the reference points

Table 6.2 shows the calculated statistical vertical errors for each different DEM for the same study area. As anticipated, LiDAR is not only the most precise DEM because of its highest resolution, but also the most accurate. The *RMSE* of each LiDAR DEMs increased from 0.58 m (Jhr Ref) to 1.27 m (Jhr L90) as the resolution of the DEMs reduced from 1 m (original resolution) to 90 m.

The *RMSE* value of the other DEMs is 4.66 m for contour maps (Jhr T20), 7.01 m for ASTER (Jhr A30) and 6.47 m for SRTM (Jhr S90). It's also noticeably that the *RMSE* of the SRTM DEM for this particular study area is within the average height accuracy found in other SRTM literature either global or at particular continent (see Table 6.3). It is proven that this type of DEM gives an acceptable result when used in large scale flood modelling (e.g. Patro *et al.*, 2009; Paiva *et al.*, 2011; Yan *et al.*, 2013).

Table 6.2: Statistics of error (m) of each DEMs with respect of the GPS control points

Model code	Min. error (m)	Max. error (m)	*RMSE* (m)	*ME* (m)
Jhr Ref	-0.59	1.00	0.58	0.35
Jhr L2	-0.64	1.38	0.58	0.36
Jhr L20	-0.83	1.83	0.68	0.42
Jhr L30	-0.93	3.98	0.79	0.48
Jhr L90	-5.46	3.73	1.27	0.49
Jhr T20	-15.38	10.55	4.66	-0.12
Jhr A30	-33.37	7.58	7.01	-4.74
Jhr S90	-3.59	4.32	6.47	-0.49

Despite having the lowest vertical accuracies, the ASTER and contour DEMs are still widely used in the field of hydraulic flood research as they are globally available and free (e.g. Tarekegn *et al.*, 2010; Wang *et al.*, 2011; Gichamo *et al.*, 2012). The differences in the vertical accuracies may partly due to the lack of information in topographical flats areas such as floodplains.

In terms of *ME* values for all LiDAR DEMs models show the positive value indicating the overestimation of the ground elevation, while the other source of DEMs (contour maps, ASTER and SRTM) indicating the opposite value.

Although LiDAR DEM gives the lowest error, it is useful to note that this type of DEM has a number of limitations as highlighted in the several papers (see Sun *et al.*, 2003; Casas *et al.*, 2006; Schumann *et al.*, 2008):

i.   it provides only discrete surface height samples and not continuous coverage,

ii. its availability is very much limited by economic constraint,

iii. its inability to capture the river bed elevations due to the fact the laser does not penetrate the water surface, and

iv. its incapability to penetrate the ground surface in densely vegetated areas especially for the tropical region.

Table 6.3: Reported vertical accuracies of SRTM data

Reference	Average height accuracy (m)	Continent
Rabus et al. (2003)	6.00	European
Sun et al. (2003)	11.20	European
SRTM mission specification (Rodriguez *et al.*, 2005)	16.00	Global
Berry et al. (2007)	2.54	Eurasia
	3.60	Global
Farr et al. (2007)	6.20	Eurasia
Wang et al. (2011)	13.80	Eurasia

Overall, the terrain is considered well defined under the LiDAR DEMs even though the calculated errors (*RMSE*) are higher compared to the vertical accuracy reported in product specification (around 0.15 m). Figure 6.2 show the distribution of each DEMs compared to the GPS ground elevation.

However, the further use of each DEM in this study is subject to its performance in the hydraulic flood modelling during the calibration and validation stages, which are described in the following sub-section.

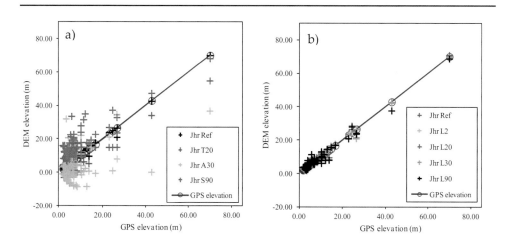

Figure 6.2: Comparison between GPS point elevations and elevations derived by the different DEMs: a) different sources of DEMs, and b) LiDAR DEMs at different resolution

6.4.2 Model calibration and validation

Figure 6.3 shows the model response in terms of *MAE* provided by the eight models (Table 6.1) in simulating the 2006 flood event. In general, all models showed to be more sensitivity to the changing of Manning's *n* roughness coefficient of main channel than the Manning's *n* roughness coefficient of floodplain areas. The results of the calibration showed that the best-fit models based on LiDAR DEM with different resolutions (Jhr L2, Jhr L20, Jhr L30 and Jhr L90) generally gave good performances with only slight variations in the *MAE* value from 0.38 m to 0.41 m.

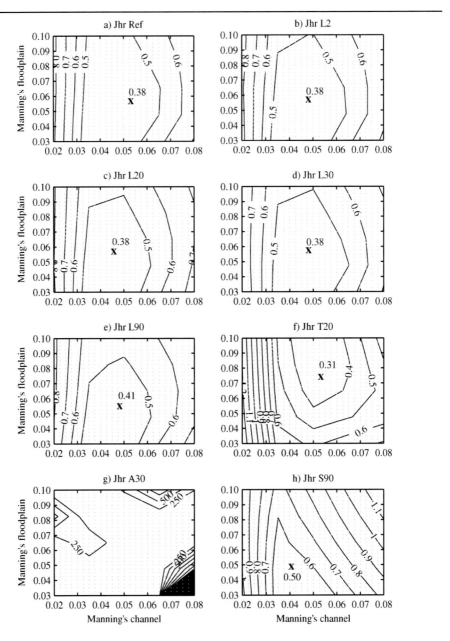

Figure 6.3: Contour map of MAE across the parameter space for twelve different models. x-axis represent Manning's n channel and y-axis represent Manning's n floodplain

Nevertheless, the optimum channel and floodplain Manning's n roughness coefficient are centred on similar values at $n_{channel}$ = 0.0425 to 0.0500 and $n_{floodplain}$ = 0.0575 for Jhr Ref, Jhr L2, Jhr L20, Jhr L30 and Jhr L90. While, the best-fit models based on topographic map and SRTM also performed well with *MAE* of 0.31 m and 0.50 m. On the other hand, ASTER-based model completely failed (exceptionally high value of *MAE* in Figure 6.3g are due to model instabilities) and was therefore eliminated from further analysis.

Table 6.4: Model validation results for different quality and accuracy of DEM models

Model name	Calibrated Manning's n roughness coefficient		MAE (m) (validation)
	Channel	**Floodplain**	
Jhr Ref	0.0500	0.0575	0.40
Jhr L2	0.0450	0.0575	0.38
Jhr L20	0.0425	0.0575	0.37
Jhr L30	0.0450	0.0575	0.38
Jhr L90	0.0450	0.0550	0.39
Jhr T20	0.0500	0.0750	0.60
Jhr S90	0.0375	0.0500	0.60

The best-fit models, using the optimum Manning's n roughness coefficients (Table 6.4), were then used to simulate the January 2007 flood event for model validation. This was carried out for all models except ASTER based model due to its poor performance (see Figure 6.3g). Table 6.4 summarises the *MAE* of each model obtained during model validation. It is noted that the *MAE* values for all LiDAR based models (Jhr L2, Jhr L20, Jhr L30 and Jhr L90) with different resolutions remained almost the same with the difference within +0.02 m. The *MAE* values for the models based on topographic contour maps and SRTM DEM provides a *MAE* of 0.60 m.

The results of this first analysis suggest that the reduction in the resolution of LiDAR

DEMs (from 1 m to 90 m) does not significantly affect the model performance.

However, the use of topographic contour maps (Jhr T20) and SRTM (Jhr S90) DEMs

as geometric input to the hydraulic model produces a slight increase of model errors.

For instance, Jhr L90 and Jhr S90 have the same resolution (90 m), but the different

accuracy results into increased (tough not remarkably) errors in model validation

(from 0.39 m to 0.60 m). This limited degradation of model performance (Table 6.4),

in spite of the much lower accuracy of topographic input (Table 6.2) can be attributed

to the fact that models are compared to water levels observed in two cross-sections.

A spatially distributed analysis (comparing the simulated flood extent and flood

water profile along the river) might show more significant differences (see Section

6.4.3).

6.4.3 Quantifying the effect of the topographic data source on the water surface elevation and inundation area

Inundation area (sensitivity analysis)

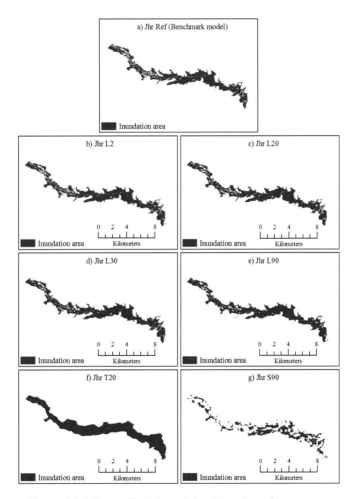

Figure 6.4: Effects of DEMs on Johor River inundation maps

This section reports an additional analysis aiming to better explore the sensitivity of model results to different topographic data. Figure 6.4 shows the simulated flood extent maps obtained from the seven different topographic input data. The floodplain areas simulated by the five LiDAR-based models (Jhr Ref, Jhr L2, Jhr L20, Jhr L30 and Jhr L90) are very similar. In contrast, the floodplain areas simulated by the models based on topographic contour maps (Jhr T20) and SRTM DEM (Jhr S90) are substantially different (see Figure 6.4 and Table 6.5).

Table 6.5: Effects of DEMs (source and resolution) on HEC-RAS simulations

Model name	Inundation area (km²)	Area difference (%)	F (%)
Jhr Ref	25.9	-	-
Jhr L2	25.8	- 0.3	96.6
Jhr L20	26.0	0.4	92.9
Jhr L30	26.2	1.2	92.2
Jhr L90	25.8	- 0.1	89.4
Jhr T20	29.2	13.0	73.7
Jhr S90	16.6	- 35.9	48.9

Furthermore, the aforementioned measure of fit F was found to decrease for both decreasing resolution and lowering accuracy. This sensitivity analysis also shows that the results of flood inundation models are more affected by the accuracy of the DEM used as topographic input than its resolution. Table 6.5 shows the comparison between the different models in terms of simulating flood extent.

Water surface elevation

Figure 6.5 compares the flood water profiles simulated by the reference model (Jhr Ref) with the flood water profiles (*WSE*) obtained from the other eleven models. All these flood water profiles were obtained by simulating the 2007 flood event. Despite having different resolutions, the flood water profiles simulated from all LiDAR-based models portray a similar flood water profiles to the reference model [see Figure 6.5(a) to Figure 6.5(d)]. This is consistent with the findings about the inundation area (Figure 6.4). Whereas, flood water profiles simulated by the models based on topographic contour maps and SRTM DEMs [see Figure 6.5(e) and Figure 6.5(f)] are rather different.

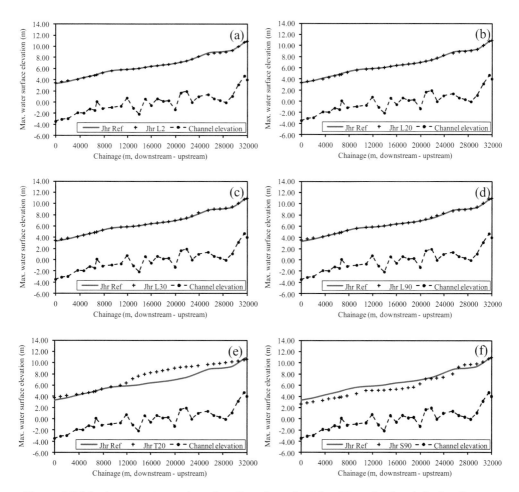

Figure 6.5: Maximum water surface elevation along the Johor River for the six hydraulic models compared to that simulated by the reference model (Jhr Ref)

The discrepancies between the reference model (Jhr Ref) and the other models visualized in Figure 6.5 are quantified in terms of Mean Absolute Error (MAE). This shows that the re-sampled LiDAR DEM models (Jhr L2, Jhr L20, Jhr L30 and Jhr L90) have a low *MAE*: between 0.05 m to 0.08 m. Higher discrepancies are found with the

models based on contour maps (1.12 m) and SRTM DEM (0.76 m). The great

differences obtained using the topographic contour maps may be partly due to the

way that the DEM height is sampled. For instance, contour DEM in this study were

based on topographic contours at 20 m intervals and required interpolation

technique to generate a DEM. Table 6.6 shows the summary of *MAE* in terms of

water surface elevation simulated by the models.

Table 6.6: Summary of Mean Absolute Error *(MAE)* in terms of water surface elevation
simulated by the models

Model name	MAE_{WSE} (m)
Jhr Ref	-
Jhr L2	0.06
Jhr L20	0.05
Jhr L30	0.05
Jhr L90	0.08
Jhr T20	1.12
Jhr S90	0.76

Uncertainty in flood profiles obtained from different DEMs model by considering parameter
uncertainty

To better interpret the differences that have emerged in comparing the results of

models based on different topographic data, a set of numerical experiments were

carried out to explore the uncertainty in model parameters. As mentioned, we varied

the Manning's *n* roughness coefficient between 0.02 and 0.08 m$^{-1/3}$s, for the river

channel, and from 0.03 to 0.10 m$^{-1/3}$s, for the floodplain, with steps 0.0025 m$^{-1/3}$s. Then,

a number of simulations are eliminated as described in Section 6.3.4.

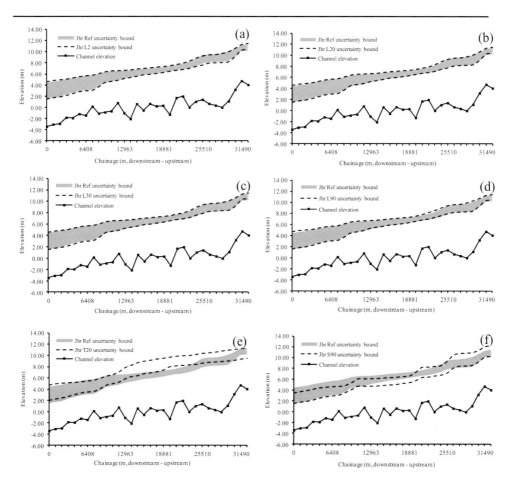

Figure 6.6: Comparison of uncertainty bounds (5th and 95th percentiles by considering parameter only) between the reference model and other models

Figure 6.6 shows the uncertainty bounds for the different models. The width of these uncertainty bounds was found to be between 1.5 m and 1.6 m for all models (only parameter uncertainty is considered here). Nevertheless, the model based on contour maps lead to significant differences from the LiDAR based model, even when the uncertainty induced by model parameters is expletively accounted for [see Figure 6.6(e)].

Table 6.7: Average value and standard deviation of the width of the uncertainty bounds estimated for all models

Model name	Mean (m)	Standard Deviation (m)
Jhr Ref	1.78	0.66
Jhr L2	1.77	0.71
Jhr L20	1.77	0.72
Jhr L30	1.77	0.71
Jhr L90	1.80	0.75
Jhr T20	1.97	0.34
Jhr S90	1.76	0.37

6.5 Conclusions

This chapter assessed how different DEMs (derived by various sources of topographic information or diverse resolutions) affect the output of hydraulic modelling. A reach of the Johor River, Malaysia, was used as the test site. The sources of DEMs were LiDAR at 1 m resolution, topographic contour maps at 20 m resolution, ASTER data at 30 m resolution, and SRTM data at 90 m resolution. The LiDAR DEM was also re-sampled from its original resolution dataset to 2, 20, 30 and 90 m cell size. Different models were built by using them as geometric input data.

The performance of the five LiDAR-based models (characterised by different resolutions ranging from 1 to 90 m; see Table 6.4) did not show significant differences. Neither in the exercise of independent calibration and validation based on water level observations in an internal cross section, nor in the sensitivity analysis of simulated flood profiles and inundation areas. Another striking result of our study is that the model based on ASTER data completely failed because of major inaccuracies of the DEM.

In contrast, the models based on SRTM data and topographic contour maps did relatively well in the validation exercise as they provided a mean absolute error of 0.6 m, which is only slightly higher the ones obtained with LiDAR-based models (all around 0.4 m). However, this outcome could be attributed to the fact that validation could only be performed by using the water level observed in a two internal cross-sections. As a matter of fact, higher discrepancies emerged when LiDAR-based models are compared to the models based on SRTM data or topographic contour maps in terms of inundation areas or flood water profiles. These differences were found to be relevant even when parameter uncertainty is accounted for.

The chapter also showed that, to support flood inundation models, the quality and accuracy of the DEM is more relevant than the resolution and precision of the DEM. For instance, the model based on the 90 m DEM obtained by re-sampling the LiDAR data performed better than model based on the 90 m DEM obtained from SRTM data. These outcomes are unavoidably associated to the specific test site, but the methodology proposed here can allow a comprehensive assessment of the impact of diverse topographic data on hydraulic modelling of floods for different rivers around the world.

Chapter 7
Uncertainty in simulating design flood profiles and inundation maps on the Johor River, Malaysia

7.1 Introduction

In hydraulic modelling of floods, estimating the potential flood extent, i.e. inundation map, and maximum water levels, i.e. flood profile, corresponding to a river discharge with a given return period, i.e. design flood, is an important step for assisting decision makers in flood risk management. A traditional representation of simulated results in flood modelling is based on a single simulation used as the best estimate. This approach, which can be called as "deterministic", does not explicitly account for the uncertainties in both the estimation of the design flood (Di Baldassarre *et al.*, 2010) and the hydraulic modelling process (Bates *et al.*, 2004) and may lead to an inaccurate hazard assessment as highlighted in the recent literature (e.g. Beven, 2009; Domeneghetti *et al.*, 2013; Dottori *et al.*, 2013). For design flood profile, this uncertainty is sometimes accounted for by adding arbitrary freeboard heights (of e.g. 1 foot or 1 meter; Brandimarte and Di Baldassarre, 2012) to the simulated maximum water levels.

Several sources of uncertainty have been shown to significantly affect the estimation of design flood profiles and inundation maps, including high flow data, hydraulic model parameters (e.g. Manning's *n* roughness coefficients) and topography data. High flow data are considered one of the most uncertain variables in hydraulic modelling of floods (Pappenberger *et al.*, 2006). As an example, Di Baldassarre and

Montanari (2009) concluded the use of rating curves to estimate river flows might lead to errors up to 40% for flood conditions. These errors propagate in the hydraulic modelling exercise and increase the uncertainty in the estimation of design flood profiles and inundation map. Moreover, the uncertainties in inflow data not only arise from the way flow data are observed, but also from the probabilistic methods used to estimate the design flood (Di Baldassarre *et al.,* 2012).

Another important source of uncertainty, which gives a significant impact in hydraulic modelling, are the friction parameters, i.e. Manning's *n* roughness coefficients. Different Manning's *n* are often used to represent the channel and floodplain friction conditions and they are treated as calibration parameters (Horritt, 2005). Although the roughness coefficient is theoretically different from one point to another, most studies use lumped values of roughness for the channel or floodplain or even the entire cross-sections (e.g. Pappenberger *et al.,* 2005).

Topography data are also significant sources of uncertainty in hydraulic modelling of floods. For instance, Schumann et al. (2008) demonstrated the effects of DEMs on deriving the water stage and inundation area. Three DEMs at three different resolutions from three sources (lidar, contour and SRTM DEM) were used for a study area in Luxembourg. By using the HEC-RAS 1-D hydraulic model to simulate the flood propagation, the result shows that the lidar DEM derived water stages by displaying the lowest RMSE, followed by the contour DEM and lastly the SRTM. In addition, uncertainty in topography affects the simulated flood extent in the process of transferring the simulated water surface elevations into inundation maps.

A number of studies have considered the relative impact of the above sources of uncertainty in the results of hydraulic models. It was found, for instance, that the overall uncertainty in design flood profiles is often affected more by the estimation of design flood values than by Manning's *n* roughness coefficients (Brandimarte and Di Baldassarre, 2013).

There are several techniques to assess uncertainty in hydraulic modelling of floods, such as Bayesian forecasting (Krzystofowich, 1999, 2002), a methodology using a fuzzy extension principle (Maskey *et al.*, 2004), parameter estimation by sequential testing (PEST) (Liu *et al.*, 2005) and generalized likelihood uncertainty estimation (GLUE) (Blazkova and Beven, 2009; Yatheenradas *et al.*, 2008). Among the several uncertainty techniques listed above, the GLUE methodology proposed by Beven and Binley (1992) is one of the most common methods to represent uncertainty in hydraulic and hydrological predictions. This technique has been widely used in various studies related to uncertainty, which include flood inundation mapping and design flood profile (Romanowich and Beven, 1998; Pappenberger *et al.*, 2005; Brandimarte and Wolyes, 2013).

The objective of this paper is to explore and quantify the difference arising between deterministic approaches and uncertainty analysis in simulating design flood profile and flood inundation maps. This study focuses on the uncertainty in hydraulic model parameters (Manning's *n* roughness coefficients) and in the estimation of the design flood (1-in-100 year flood hydrograph), as both uncertainties have shown to be significant in the application of 1-D hydraulic modelling (Pappenberger *et al.*, 2008; Di Baldassarre *et al.*, 2010; Brandimarte and Woldeyes, 2013). The topographic uncertainty was neglected in this research, which is supported with high quality topographic data of Digital Elevation Model (DEM), LiDAR (Light Detection and Ranging). Yet, it is important to note that the impact of topographic uncertainty in hydraulic modelling of floods has been recently discussed in Md Ali *et al.*, (2015).

7.2 Methodology

This research used the 1-D code HEC-RAS (US Army Corps of Engineers, Hydrologic Engineering Center, 2001) for simulating the hydraulic behaviour of the Johor River between Rantau Panjang and 5 km downstream of Kota Tinggi (Fig. 1), in unsteady flow conditions. The unsteady numerical code implemented in the HEC-RAS model solves the St Venant equations for a 1-D schematization through an algorithm that uses a classical implicit four-point finite difference scheme. Despite the recent development of availability to obtain two-dimensional (2-D) hydraulic models in hydraulic modelling of floods, the 1-D hydraulic model is still widely used. Moreover, a number of studies show that 1-D hydraulic model is reliable in simulating flood propagation in natural rivers (e.g. Horrit and Bates, 2002; Castellarin et al., 2009). A friction slope at the downstream cross-section was used as downstream boundary condition, while the observed flow hydrograph at an hourly time step was used as upstream boundary condition.

7.2.1 Model calibration and validation

Two sets of discharge data from the two flood event (e.g. December 2006 and January 2007 flood event) were used as inflow data. The measured peak flow of the December 2006 flood event, with a peak flow of ~375m^3s^{-1} was used in the calibration. Whereas, the discharge data from the January 2007 flood event, with a peak flow of ~595m^3s^{-1} was used for the validation of the calibration model.

The model was calibrated by varying the Manning's n roughness coefficient against the simulated and observed water elevation after the December 2006 flood event. To assess the sensitivity of the different models to the model parameters, the Manning's n roughness coefficients for all the models were sampled uniformly from 0.02 to 0.08

m$^{-1/3}$s for the river channel, and between 0.03 to 0.10 m$^{-1/3}$s for the floodplain. The range of Manning's *n* roughness coefficient that were used in this research were chosen from the documented ranges roughness parameters (Chow 1959), and it is sufficiently large to cover the possible combinations for hydraulic modelling. Then, the performance of the hydraulic model in producing the December 2006 flood profile was assessed by means of the Mean Absolute Error (*MAE*).

7.2.2   Estimation of design flood profile

The current approach of estimating safety level of the flood protection structure (e.g. dykes, levees) is by considering a freeboard height with the design flood profile. Whereas, the normal approach to determine the design flood profile is by simulating the calibrated hydraulic model with the design flood hydrograph. In Malaysia, DID highlighted in the manual handbook that the minimum freeboard height to consider when designing the flood protection structure is 1 meter above the 1-in-100 year design flood profile. Thus, for the specific application example, the 'best fit' was run in unsteady conditions using the estimated Q_{100} at upstream boundary condition (i.e. Rantau Panjang; see Figure 4.1).

However, several scientific literatures concern that the use of the 'best fit' model with single design flood estimation in prediction of design flood profiles might misrepresent the existence of any sources of uncertainties in hydraulic modelling simulation (i.e. Beven and Freer, 2001; Beven, 2006; Di Baldassarre *et al.*, 2010; Yan *et al.*, 2013). Therefore, to avoid overlooking the existence of any sources of uncertainties which might affect the simulated design flood profile, the use of the probabilistic approach is recommended.

A number of quantitative analyses of uncertainty in hydraulic modelling have been previously reported, but these studies typically focused on only one source of uncertainty, such as the friction coefficient (Aronica *et al.*, 2002; Bates *et al.*, 2004), grid cell size (Werner, 2001), the quality of the DEM used (Casas *et al.*, 2006) or flow characteristic (Purvis *et al.*, 2008). This is mainly attributed to the fact that conducting a detail analysis that considers all source of uncertainty in the estimation of the design flood profile is not a simple task. The work would be computationally infeasible and it would require strong assumptions on the nature of errors. In addition, the data required to apply rigorous methods of uncertainty analysis are seldomly available (Beven, 2006).

Estimation of design flood profile with uncertainty

As mentioned above, Manning's *n* roughness coefficient was selected as the first variable of uncertainty in hydraulic modelling to be considered in the research. The model was run in a Monte Carlo framework to assess the parameter uncertainty using GLUE. For this scenario, given the accuracy of the used calibration data, all the couples of Manning's *n* roughness coefficient (river channel and floodplain) giving a *MAE* larger than 1-m are rejected. This choice of the rejection criteria for the *MAE* is against the requirement made by DID which determined that the minimum freeboard of any flood defence structure (i.e. dykes, levees) for main river should not be less than 1-m above 1-in-100 years flood event. In addition, the Manning's *n* roughness coefficients on the floodplain which are smaller than the channel are also eliminated. By adopting the GLUE methodology framework, all the models which passed these rejection criteria are considered as 'behavioural', and are used to simulate 1-in-100 years flood event.

According to the GLUE framework (Beven and Binley, 1992), each simulation, i, is associated to the (generalized) likelihood weight, W_i, ranging from 0 to 1. The weight, W_i is expressed as a function of the measure fit, ε_i, of the behavioural models.

$$W_i = \frac{\varepsilon_{max} - \varepsilon_i}{\varepsilon_{max} - \varepsilon_{min}} \tag{7.1}$$

where, ε_{max} and ε_{min} are the maximum and minimum value of *MAE* of behavioural models. Then, the likelihood weights are the cumulative sum of 1 and the weighted 5th, 50th and 95th percentiles. The likelihood weights were calculated as follow:

$$L_i = \frac{W_i}{\sum\limits_{i=1}^{n} W_i} \tag{7.2}$$

Although some aspects of this methodology are criticized in several papers (e.g. Hunter *et al.*, 2005; Mantovan and Todini, 2006; Montanari 2005; Stedinger *et al.*, 2008), it is still widely used in hydrological modelling because of its easiness in implementation and a common-sense approach to use only a set of the "best fit" models for uncertainty analysis (e.g. Shrestha *et al.*, 2009; Vázquez *et al.*, 2009; Krueger *et al.*, 2010).

The second variable of uncertainty in this research is the uncertainty in design flood induced by data error. Given that our research aims to evaluate two different approaches of design policies, the analysis was limited to hydraulic modelling only. The detailed analysis of derivation of the 1-in-100 year design flood hydrograph was not investigated. This scenario however referred instead to 1-in-100 year design flood hydrograph at Rantau Panjang identified from previous research conducted by DID (2009) (see Figure 7.1a).

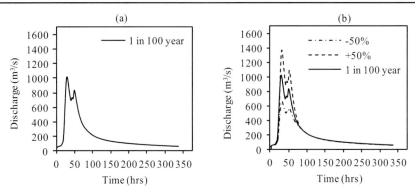

Figure 7.1: Design flood hydrograph (a) 1-in-100 year design flood hydrograph (b) 1-in-100 year design flood hydrograph by considering of data error

To evaluate the impact of data error into design flood, we generated 200 values of 1-in-100 years design flood hydrograph, uniformly distributed, by adopting the Equation 6.3 and 6.4 proposed by Kuczera (1996) and Di Baldassarre et al. (2012). According to Kuczera (1996) the systematic data error can be described as

$$Q = Q' \text{ if } \qquad Q' < Q_a \tag{7.3}$$

$$Q = Q_a + \alpha\left(Q' - Q_a\right) \text{ if } Q' > Q_a \tag{7.4}$$

Where Q indicates the observed value, Q' refers to the true value of river discharge, Q_a represents the river discharge value overspill in times of high rainfall, and α is a positive value coefficient. Here, the Q_a value is 300 m^3s^{-1} and $\alpha = \pm 50\%$. Figure 7.1b shows the results obtained with $\alpha = 0.50$ (dash dot line) and $\alpha = 1.50$ (dash line) and allows an initial interpretation of the practical effects induced by the extrapolation error.

By combining the uncertainty in the Manning's n roughness coefficient and estimation of design flood profile, the combined uncertainty was assessed. A total of 4,800 simulations were carried out by feeding all the behavioural models from first scenario (Manning's *n* roughness coefficient) with 200 discharge values from second scenario (estimation of design flood profile).

7.2.3 Simulation of flood inundation maps

To generate the deterministic 1-in-100 year flood inundation map, a 1-in-100 year design flood hydrograph was used as an input (i.e. boundary condition) with the 'best fit' model. The 'best fit' model was selected from the calibration process (see Section 7.2.1). By integration in a GIS environment, the result of the 'best fit' model was then presented in term of 1-in100 year flood inundation map.

Simulation of flood inundation maps with uncertainty

In the studies of extent validation, GLUE methodology dictates that the actual flood event's water boundaries and its performance measurement must be presence in order to estimate the Monte Carlo ensemble of likeliness of the predicted outputs against the observed data (Horritt 2006). A particular measurement of likelihood used in the calculation of the extents of forecasted flood event is presented in Horrit et al. (2007) and Di Baldassarre et al. (2009) as:

$$F = \frac{A}{A + B + C} \qquad\qquad (7.5)$$

Wherein *A* refers to the size of area of both the actual and simulated data that is overlapped; *B* is the area that is supposedly dry but inundated in the calibration; and, *C* is the area in the model that supposed to be flooded, but not. *A*, *B* and *C* denotes as the numbers of pixels correctly predicted as wet and dry in reference and predicted inundation maps. A dry and wet pixels are represented by code, either as value 0 or 1. The closer the derivation value of *F* is to 1, the more likely the outcome of the model output is comparable to the actual flood extents. Whereas if the value closer to

0, the poorer the likelihood of the output to the actual flood extents. For the study, in the absence of any available flood inundation map for Johor River, a flood inundation map for the 1-in-100 year design flow flood hydrograph was used as reference.

At this stage, all the behavioural models from the above were used to simulate the 1-in-100 year flood event. Then, the likelihood weights, L_i were rescaled to a cumulative sum of 1 and the weighted 5th, 50th and 95th percentiles, according to the values of F. The likelihood weight, L_i is expressed as a function of:

$$L_i = \frac{F_i - (F_i)_{min}}{(F_i)_{max} - (F_i)_{min}} \qquad (7.6)$$

Where $(F_i)_{min}$ and $(F_i)_{max}$ are the minimum and maximum values of F_i. In order to produce an additional uncertain 1-in-100 year flood inundation map, by considering the uncertainty in the design flood estimation, a simplified approach was applied by generating the 200 values of 1-in-100 years design flood hydrograph with all behavioural models.

7.3 Results and discussion

7.3.1 Calibration and validation

Figure 7.2 shows the model response in terms of *MAE* for the December 2006 flood event. There are different sets of parameters that provide a *MAE* lower than 0.4 m, which is relatively a good performance. The best-fit model performing combination of channel and floodplain coefficients lie inside the hyperbolic shape area by a *MAE* value of 0.38 m. In addition, the model shows to be more sensitive to the Manning's *n* channel roughness coefficient than to the Manning's *n* floodplain roughness coefficient. The optimum Manning's *n* roughness coefficients from the best-fit model above were then used to simulate the January 2007 flood event for model validation. It is found that the *MAE* value for the validation is 0.44 m.

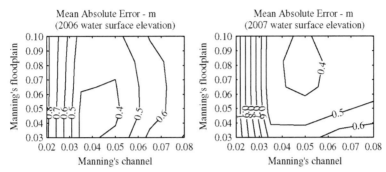

Figure 7.2: Model responses to variations in Manning's *n* roughness coefficients: (left panel) 2006 flood event; (right panel) 2007 flood event

7.3.2 Estimation of design flood profile

Figure 7.3a shows the result of simulation in terms of water level of 1-in-100 year flood profile and the profile of water level with a 1-m freeboard. Here, the profile of water level with a 1-m freeboard will be made as a reference height for safety level of any flood protection structure.

For the design flood profile, the 1-in-100 year design flood profile for this research is based on the assumption that the hydraulic model calibrated on the 2006 flood event is able to predict the 1-in-100 year flood profile. In addition, a 1-m freeboard height is added to design flood profile with the aims to compensate the many unknown factors that could contribute to flood heights greater than the height calculated for a selected size flood.

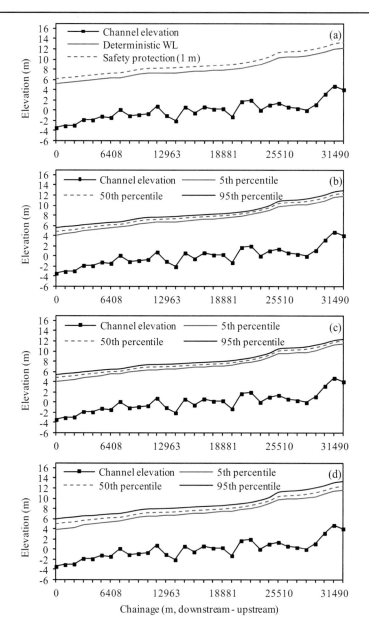

Figure 7.3: Design flood profiles: (a) design flood profile base on single simulation with best fit model and 1-m freeboard; (b)-(d) show uncertain design flood profiles by considering the (b) uncertainty in model parameter only; (c) uncertainty in design flow only; and (d) combined uncertainty between model parameter and design flow

However, the use of freeboard definition in this approach may lead to wrong assumption should the design flood structure (i.e. levee) which is adopting 1-in-100 year design flood profile with 1-m freeboard be assumed as able to protect the flood prone area from extreme flood events of more than 1-in-100 year flood. This assumption have to be corrected by understanding that the freeboard does not provide an additional safety level to the flood prone area but rather to account for the overall uncertainty which may not have been considered during the hydraulic modelling.

Table 7.1: Results of simulation for estimation of design flood profile

	Average WSE (m) from change in		
	Roughness	Inflow	Combined
Min	7.41	7.20	7.13
Max	8.67	8.25	9.21
(Max – Min)	1.26	1.05	2.08
5th percentile	7.51	7.26	7.39
50th percentile	8.07	7.80	8.14
95th percentile	8.57	8.21	8.81

Results from Monte Carlo simulations for each uncertainty variables (i.e. Manning's *n* roughness coefficient, flow and combined) when estimating design flood profile are presented in Table 7.1 and Figure 7.3b, 7.3c and 7.3d. The statistics presented in the Table 7.1 were computed using the data from all simulations for a particular variable or their combination. For example, a minimum *WSE* of 7.13 m from combined uncertainty (Table 7.1 first row, fourth column) represents the minimum *WSE* among all cross-sections from all 4,800 simulations. The *WSE* and uncertainty bounds in Figure 7.3b, 7.3c and 7.3d are also computed using the data at each cross-section from all simulations for specific variable, or the combination of the two variables. Combined uncertainties produced the largest deviation in both minimum

115

and maximum *WSE* by reducing the minimum *WSE* by 0.07 m and 0.28 m, while increasing the maximum *WSE* by 0.24 m and 0.60 m. In this research, uncertainty in Manning's *n* roughness coefficient creates a wider deviation with uncertainty bound of 1.06 m compared to inflow. Combining the uncertainty in Manning's *n* roughness coefficient and inflow data further enlarge the width of the uncertainty bound to 1.42 m.

However, for this specific research, it was found that the design flood profile with 1-m freeboard is higher than the 95th percentile uncertainty bound for each uncertainty variables either individual or combined. Hence, the standard freeboard overestimates the overall uncertainty.

7.3.3 Simulation of flood inundation map

Simulation of flood inundation map with deterministic approach

As shown in Figure 7.4, the best fit HEC-RAS model is characterised by an MAE value equal to 0.38 (optimum Manning's *n* channel = 0.04 $m^{-1/3}s$ and Manning's *n* floodplain = 0.05 $m^{-1/3}s$). A 1-in-100 year design flow hydrograph with a peak discharge ~1018 m^3/s as estimated by DID (2009) was used as a boundary condition. This best fit model was then used to simulate the deterministic 1-in-100 year flood inundation map as shown in Figure 7.4. However, the estimation of the deterministic 1-in-100 year flood inundation map was inevitably affected by several sources of uncertainty. The following section will further elaborate estimation of flood inundation map with uncertainty.

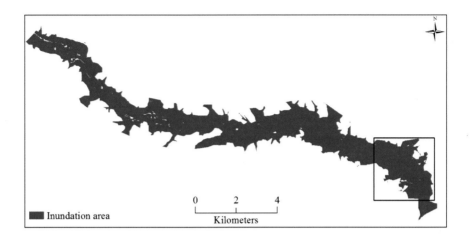

Figure 7.4: Black box indicated an enlarger area in Figure 7.5

Estimation of flood inundation map with uncertainty

Table 7.2 shows the 5th and 95th percentile of uncertainty in flood inundation area from individual effects of Manning's *n* roughness coefficient and inflow data for the study area. The uncertainty of inundation area is in the range of 2.34 km² for Manning's *n* variable, while 1.78 km² for error in inflow variable. At individual variable level, changing in Manning's *n* parameter produced the widest difference in inundation area for Johor River.

However, the difference of inundation area between the 95th percentile (for both Manning's *n* roughness = 28.10 km² and inflow = 27.88 km²) with deterministic approach (27.83 km²) is relatively small. This is partly contributed from the HEC-GeoRAS domain where the inundation extent follows the width of the cross-sections which creates a bounding polygon of flood extent.

Table 7.2: Simulation results for Johor River

	Flood inundation area (km²) from changes in	
	Manning's *n*	Flow
Min	25.41	24.94
Max	28.43	27.94
(Max – Min)	3.02	3.00
5th percentile	25.76	26.10
95th percentile	28.10	27.88

As an additional, the comparison between deterministic and probabilistic flood inundation maps which may affect flood management are also being considered in this research. Figure 7.5 compares different flood inundation maps i.e. deterministic flood inundation map (Figure 7.5a), probabilistic flood inundation maps computed on the basis of the uncertainty in model parameter (Figure 7.5b), and uncertainty in inflow data (Figure 7.5c). The probabilistic map in Figure 7.5b shows the difference in probability of inundation arising from the consideration of the uncertainty in model parameters compared to the probabilistic map in Figure 7.5c which considered uncertainty in inflow data. Areas highlighted in the figures emphasis the difference between the two variables and these show a possible misinterpretation of hazard estimation.

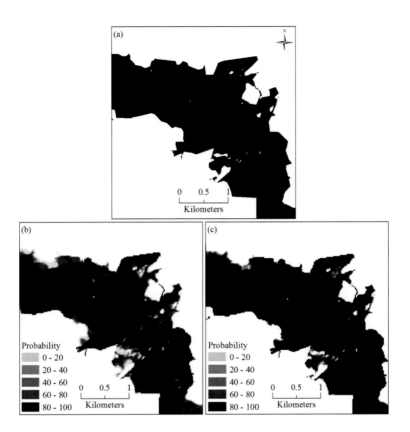

Figure 7.5: (a): Deterministic inundation map 1-in-100 year; (b) probabilistic inundation map for uncertainty in model parameter; and (c) probabilistic inundation map for uncertainty in inflow data

Although the information of potentially flooded area may also be obtained from a deterministic inundation map (see Figure 7.5a), representing the uncertainty of the output as the added value in a probabilistic inundation maps (see Figure 7.5b and Figure 7.5c) justifies the necessity for scientist to interact with decision-makers, who may or may not have a good knowledge in hydraulic modelling. Unfortunately, scientific community rarely prefers to adopt the probabilistic approaches in this field of research (Pappenberger and Beven, 2006; Di Baldassarre *et al.*, 2010). For instance, the public community would find the deterministic inundation map (see Figure 7.5a) easier to understand compared to the probabilistic map.

Furthermore, there are no guidelines in defining or quantifying the uncertainties that are used for data collection, model simulation and creating flood inundation maps. For example, is the result less than 5th uncertainty bound is acceptable? Thus, to promote the usage of probabilistic maps among the decision-makers, a modeller should provide the supporting information on the interpretation of distinction between the overall probabilities and those associated with respective uncertainty scenarios.

7.4 Conclusions

The accuracy in developing flood inundation map and prediction of design flood profile are important in flood hazard management. However, the accuracies of the flood map and flood profile depend on the uncertainty of the variables used in the flood modelling process. Although a modeller may knows the uncertainties in data and model parameters during the hydraulic modelling process, understanding and quantifying the role of each uncertain variable in the production of flood inundation map and prediction of design flood profile are still complicated.

The following conclusions are drawn from this specific case research:

- The uncertainty from the Manning's n roughness coefficient has greater influence to flood inundation map than the uncertainty from inflow data. This finding confirms the importance of this variable in the overall flood modelling process, but contrasts with previous studies (e.g. Brandimarte and Di Baldassarre, 2012; Domeneghetti *et al.*, 2013), which found a larger impact of the uncertainty in inflow data.

- The use of a 1-D hydraulic model and its associated assumptions need additional work. A 1-D model such as HEC-RAS provides water surface elevations only along individual cross-sections. Uncertainty can arise in this approach when the water surface elevations at cross-sections are interpolated to create inundation maps.

- The use of a freeboard should not be considered as an additional safety factor, but rather as a safety margin that allows for uncertainties. The finding from this research shows that the design flood profile with 1-m freeboard are higher than the 95th percentile bound for each uncertainty variables, but this cannot be generalized. The uncertainty approach proposed here allows a quantification of uncertainty based on transparent (though unavoidably subjective) assumptions.

This difference in flood inundation translates into significant underestimation of lives in jeopardy in populated area. It's also has significant implications in the estimation of flood inundation area and potentially on the floodplain emergency management plan. As mentioned above, this research involved in the use of only one particular research area, and therefore similar research must be carried out for other research areas with different topography, hydrology and sizes. Furthermore, each variable must also be investigated in more detailed with consideration of several other aspects. For instance, the flow data used in this research involved uncertainty arising by the equations. However, in reality when hydrological analyses were carried out to obtain the design flow, uncertainty from input data and hydrologic analysis should also be considered. The overall probability of inundation area within the floodplain area of Johor River could differ if all possible uncertainties from respective scenarios were considered.

Chapter 8
Conclusions and Recommendations

8.1 Conclusion

The impacts of flooding are devastating in terms of human displacements, loss of life and damage in infrastructure and property. One of the efforts to minimize losses is providing early information to the community about risks through flood inundation maps (i.e. flood hazard map). These maps do not only identify future flood prone areas, but also provide useful information for rescue and relief agency, land planners and local authority.

Developing a flood inundation map involves a series of procedure which require hydrological studies (i.e. to estimate the return period of flood event) and hydraulic modelling (i.e. for estimation of water surface elevations and inundation area).

In normal practise, the result of flood inundation modelling is based on a deterministic approach (i.e. single simulation) without considering some uncertainty, regardless of the way the maps was created. However, without considering the uncertainties (e.g. model parameters, the terrain data and model structure), the result of the flood inundation modelling may not be accurate and this may lead to wrong information given on hazard assessments.

Uncertainties in flood inundation are generated from various sources. Among others are estimation of design flow, terrain data sets (i.e. type of DEMs, source of DEMs, spatial resolution, etc.), geometric description in hydraulic models (i.e. inclusion or exclusion of the hydraulic structures, number and spacing of cross-sections, etc.) and hydraulic modelling approach (i.e. 1-D/2-D). For instance, ample researches especially in Europe has exhibit the advantages of using two-dimensional (2-D) hydraulic models for flood inundation mapping over one-dimensional (1-D) in simulating flow dynamics in floodplains (Horrit and Bates, 2002; Cobby *et al.*, 2003; Horrit *et al.*, 2006; Tayefi *et al.*, 2007).

Nevertheless, there is still a doubt regarding the selection of the right modelling approach (either 1-D or 2-D) in prediction of accurate flood inundation extent when other factors such as the availability of high accuracy topography are limited. Although with availability of detailed and accurate topographic data from LiDAR in the floodplain, the geometric description of the topography (such as cross-section spacing and resolution) on a hydraulic model can still gives a significant impact to the model output.

This thesis intends to contribute further in the knowledge of how uncertainty affects the flood inundation maps. The case study in this thesis is the Johor river reach in Johor, Malaysia, which is an agricultural-dominated area. It was chosen because of its comprehensive data available. Another unique aspect of this thesis is that the performance of the hydraulic models is evaluated using independent calibration and validation data of an observed flood water levels.

However, the intend of the thesis is not to certify any process, or to nullify the current state of process in creating flood hazard map, but rather to emphasize the impact of uncertainties in these maps. The next paragraphs will summarize the approach taken within this thesis and highlight the main conclusion of the research.

8.1.1 Summary of contributions

Research question No. 1: How do the many sources of uncertainty (e.g. hydrologic data, topographic data, and model selection) affect flood hazard mapping? This is addressed in Chapter 4 and Chapter 5 of this thesis.

Chapter 4: Cross-sections spacing in 1-D hydraulic modelling.

In illustrating the inundation extent using 1-D hydraulic model, all detail of geometric description including spacing of cross-sections, number and structural details (i.e. bridge) play an important role in presenting accurate results. In the case of Johor River, five hydraulic models were set up with different cross-sections spacing. The model set-up were based on the followings, the first model based on the original configuration of cross-sections which include the bridge detailing; the next two models based on modeller own judgement (by reducing the numbers and spacing of cross-sections from original configuration); and the last two hydraulic models based on equation proposed by Samuels (1990) to determine the cross-sections spacing.

For Johor River, reducing the numbers and spacing of cross-sections give minor differences in terms of models performance among the hydraulic models. As an example, for the validation of model performance, a hydraulic model with higher numbers of cross-sections gives a value of *MAE* equal to 0.46 m; compared to *MAE* of 0.49 m produced by a hydraulic model with lower numbers of cross-sections. The most interesting part is that, the hydraulic model without the bridge detailing but remains the same cross-section spacing as the original configurations performed equally better as per the hydraulic model with a bridge detailing. However, reduction in the number of cross sections (increased in the spacing), does not have significant impact on the water surface profiles results. The reduction caused lack of topographical information available during the computation stage to generate accurate water surface profiles. This is agreeable with Castellarin et al. (2009) that suggested that if the purpose of hydraulic modelling is for design verification for flood mitigation structures, then detailed topographical survey data shall be required. Furthermore, increasing the spacing of the cross-sections also resulted in an increase in the inundation area. The maximum change in inundation area between the hydraulic model with the highest and the lowest numbers cross-sections is about 28%.

Chapter 5: 2-D hydraulic modelling: The role of digital elevation models.

In hydraulic flood modelling, important input data is represented in the topographical dataset of floodplain and channel. With high resolution DEMs such as LiDAR becoming more readily available, more precise results can be produced such as in predicting flood inundation extent. However, the computational cost increases as the DEMs resolution becomes finer. Furthermore, computing hardware capabilities is still a limiting factor, more so for large research area. Therefore, there is a need for the DEMs to be aggregated to a coarser resolution. With the availability of GIS software, aggregation process become easier, but without understanding of the aggregation process of DEMs, there is a possibility of loss of determining data that could compromise the accuracy and reliability of the results.

To address this, the LISFLOOD-FP model integrated with ARC View to examine how sensitive the LISFLOOD-FP model was to the different aggregation technique and resolution while predicting the performance of the model and inundation area. Based on the algorithm available in GIS, two types of DEMs were developed from three different aggregation algorithm function (mean, maximum and minimum). The first DEM, the function of min was specified for channel, and floodplain as function of max, whereas for the second DEM, both channel and floodplain DEMs was specified as function of mean. Then, the selected aggregation algorithms are used to aggregate the LiDAR DEMs from 1 m to 30, 90 and 120 m in order to simulate the model performance and inundation area of the research area.

The model predicted water level was calibrated and validated against water level gauge data. Results indicate that the re-sampled DEM (30, 90, and 120 m) from different aggregation algorithm function give significant differences in calibration and validation analyses. In term or inundation area, the results from LISFLOOD-FP show that the different of aggregation algorithm function give significant differences between the models. The result from the analysis clearly showed that by selecting the proper aggregation function will provide much better accuracy of model results.

Research question No. 2: What are the potentials and limitations of different data sources (including remote sensing) in supporting flood inundation modelling? This is addressed in Chapter 6 of this thesis.

Chapter 6: The role of digital elevation models (DEMs) in 1-D hydraulic modelling.

Sets of 1-D hydraulic model developed using different sources and resolutions of DEMs were carried out to investigate its impact on the model output. This is accomplished by using four source of DEMs dataset with different resolution: 1 m LiDAR data set; 20 m topographic dataset; 30 m ASTER DEM; and 90 m SRTM dataset. In addition, to analyze the effect of spatial resolution to model output, four additional DEMs were acquired by re-scaling the original LiDAR DEM (1 m) to the spatial resolution of 2; 20; 30; and 90 m. The impact of the DEM from different quality (i.e. accuracy) and precision (i.e. spatial resolution) was analysed in term of: (i) DEMs quality; (ii) model performance; (iii) water surface elevation and inundation area; and (iv) uncertainty analysis.

For the first analyses, in comparison between the observation control points with the topographic data set, the ASTER DEM was the least accurate with RMSE of 7.01 m. While for the topographic dataset and SRTM DEM, the RMSE is 4.66 m and 6.47 m. The results also show a slight increment of error (RMSE) of LiDAR DEM from 0.58 m to 1.27 m as the reduction of spatial resolution from 1 m to 90 m. Although the ASTER and SRTM DEMs show a higher value of error, yet it's still widely used due to it freely and globally availability.

For the calibration and validation analyses, the reduction of LiDAR DEMs resolution does not give significant changes to the model performance with different of MAE between the hydraulics model in between 0.01 m and 0.02 m. While for the model based on topographic dataset and SRTM, the differences are 0.29 m and 0.10 m. Nevertheless, the further used of ASTER DEM in the hydraulic modelling of flood were unforeseen due to the unexceptional high results during the calibration stage.

The model based on SRTM DEMs produced the lowest prediction of inundation area compared to the benchmarking results. The inundation area is 16.6 km² with measure of fit, F, 48.9 %. While for the model based on topographic dataset produce the higher error of WSE compared to benchmarking model with MAE of 1.12 m. Moreover, the entire model based with LiDAR DEMs produced the results (i.e. inundation area and water surface elevation) which are equally same as the benchmarking model. In addition, analyses of the uncertainty using GLUE methodology shows that the model based on topographic dataset provide significant differences from the benchmarking model.

Research question No. 3: How can we model uncertainty to better define the safety level of flood protection structures? This is addressed in Chapter 7 of this thesis.

Chapter 7: Uncertainty in simulating design flood profiles and inundation maps on the Johor River, Malaysia.

Estimating of flood profiles and flood inundation maps is one of key factor in disseminate/conveying flood hazard information to the public and authority (planners, emergency and relief operations). Such simulated results of flood modelling (e.g. design flood profile and inundation maps) been produced based on deterministic approaches (Bates et al., 2004; Merz et al., 2007; Di Baldassare et al., 2009). This approach is accomplished by setting up a flood hydraulic model (1-D/2-D), calibrating the model using past flood event and using the best fit model to simulate design flood profile and visualization of flood hazard from the model results in a GIS environment. Nevertheless, deterministic models may provide inaccurate results because it does only delineate a flood inundation area without considering of uncertainties from all the variables involved in the overall process of the flood modelling.

This delineation may provide misleading information to the public or emergency relief agency that they are within the hazard area or not. By considering the uncertainty variables in flood hydraulic modelling, more information may provide and could be used to guide mitigation toward to higher risk area instead of all exposed area.

Results of design flood profile show that the uncertainty in the roughness coefficient, inflow and combine of both created different uncertainty bound. For roughness coefficient, the uncertainty bound is 1.06 m compare to inflow is 0.95 m. However, when combines the two parameter, the uncertainty bound is 1.42 m. In this research, the estimated design flood profile with 1 m freeboard is higher than 95th uncertainty percentile for all parameter (either individually or combine). In addition, for flood inundation map, by considering the uncertainty in model parameter, the different of area is 2.34 km^2 where as for inflow is 1.78 km^2.

8.1.2 Recommendations

Information on flood inundation area is crucial in flood risk management. Not only important to rescue and relief agency during floods, but also to the planner when evaluate the propose land development in floodplain zone. Creating flood inundation maps involves hydrologic and hydraulic modelling; and topographical terrain analyses. However, this process is affected amongst other by input data, type of model used and river geometry in the model.

The following recommendations indicate some of the aspects of improving flood inundation modelling for further research: -

o Researcher should include the hydrological analyses such as estimation of design flow. Depending on the availability of the historical stream flow data, the approach of estimating design flow would be different. Furthermore, this research is not considering any discharge from other tributaries along the river reach.

o To further evaluate the effects of geometric data in 1-D hydraulic modelling, more variability of inclusion and exclusion of cross-section (e.g. at the river bend, width of cross-sections) should consider. Moreover, the characteristic of the river (e.g. length of the river, slope, and geomorphology) and varying upstream/downstream boundary conditions (e.g. releasing water from a dam, tide level, barrage) might give different results. In addition, research should be done in other river reach/catchment with different land use characteristics where the upstream of this research area is dominated by agriculture.

o Another area to explore is the uncertainty analyses techniques. In this thesis the GLUE methodology were adopted to perform the uncertainty analysis. Although these approaches have been debated in certain scientific research, in contrast it still widely used due to it easiness in implementation and computational efficiency. Perhaps, by using different approaches of uncertainty analyses (i.e. MLUE, UNEC, etc.) may give more precise of model outputs and understanding of uncertainty in flood modelling.

o In this thesis, the re-sampling of the DEMs is using the functions available in ArcGIS software (i.e. Spatial Analysis tools). However, there are also other functions (i.e. Data Management tools) which are not explored since these functions are commonly used for re-sampling of DEMs. Anyhow, to identify the suitable re-sampling techniques of DEMs, the used of other software such as Global Mapper should not be ignored. Therefore, the selections of the function for re-sampling or software use should considered and understand carefully.

o Further research should be done by consider other variables for calibration and validation process. In this research, only two variables were used (discharge and water level), which are sufficient for calibrating 1-D model. In contrast, it may not significant to calibrate 2-D model. To produce reliable estimation of models output and fully exploit the potential of the hydraulic, it vitals to have additional variables measurements (i.e. additional observation station along the river reach) for proper calibration process. Moreover, with advances in technology, the availability of synthetic aperture radar (SAR) images capture the flood inundation area, added a new techniques of calibration process.

o In developing the flood inundation map, the accuracy is important. However, it is dependent on the precision of the variables used (estimation of design flow, data collection techniques of DEMs, hydraulic modelling approach). With regards of the uncertain variables, the approach of using the probabilistic inundation map should be deliberate and explored further.

References

Akbari, G. H., and Barati, R. (2012), Comprehensive analysis of flooding in unmanaged catchments. *Proceedings of the Institution of Civil Engineers-Water Management*, 165(4), 229-238.

Apel, H., Thieken A. H., Merz, B. and Blöschl, G. (2004), Flood risk assessment and associated uncertainty. *Natural Hazards and Earth System Sciences*, 4, 295–308.

Aronica, G., Bates, P. D., and Horritt, M. S. (2002), Assessing the uncertainty indistributed model predictions using observed binary pattern information within GLUE. *Hydrological Processes*, 16, 2001–2016.

Asselman, N., Bates, P., Woodhead, S., Fewtrell, T., Soares-Frazão, S., Zech, Y., Velickovic, M., de Wit, A., ter Maat, J., Verhoeven, G., and Lhomme, J. (2009), Flood inundation modelling: Model choice and proper application, *Floodsite, Report-T08-09-03*.

Bales, J. D., and Wagner, C. R. (2009), Source of uncertainty in flood inundation maps. *Journal of Flood Risk Management*, 2, 139-147.

Baltsavias, E. P. (1999), A comparison between photogrammetry and laser scanning. *ISPRS Journal of Photogrammetry & Remote Sensing*, 54, 83-94.

Baptist M. J., Penning, W. E., Duel, H., Smits A. J. M., Geerling, G. W., Van der Lee, G. E. M., and van Alphen, J. S. L (2004), Assessment of the effects of cyclic floodplain rejuvenation on flood levels and biodiversity along the Rhine River. *River Research and Applications*, 20, 285–297.

Bates, P. D. (2004), Remote sensing and flood inundation modelling. *Hydrological Processes*, 18, 2593-2597.

Bates, P. D. (2012), Integrating remote sensing data with flood inundation models: how far have we got? *Hydrological Processes, 26,* 2515-2521. DOI: 10.1002/hyp.9374.

Bates, P. D., and De Roo, A. P. J. (2000), A simple raster-based model for flood inundation simulation, *Journal of Hydrology,* 236, 54–77.

Bates, P. D., Horrit, M. S., Aronica, G., and Beven, K.J. (2004), Bayesian updating of flood inundation likelihoods conditioned on flood extent data. *Hydrological Processes,* 18(17), 3347–3370.

Bates, P. D., Horritt, M. S., and Fewtrell, T. J. (2010), A simple inertial formulation of the shallow water equations for efficient two-dimensional flood inundation modelling. *Journal of Hydrology,* 387, 33–45, doi:10.1016/j.jhydrol.2010.03.027.

Bates, P. D., Marks, K. J., and Horritt, M.S. (2003), Optimal use of high-resolution topographic data in flood inundation models. *Hydrological Processes,* 17,537-557.

Berry, P. A. M., Garlick, J. D., and Smith, R. G. (2007), Near-global validation of the SRTM DEM using satellite radar altimetry. *Remote Sensing of Environment,* 106(1), 17–27. DOI:10.1016/j.rse.2006.07.011.

Berz, G., (2000), Flood disasters: lessons from the past-worries for the future. *Proceedings of the Institution of Civil Engineers-Water and Maritime Engineering,* 142(1), 3-8.

Beven, K. J. (2006), A manifesto for the equifinality thesis. *Journal of Hydrology,* 320, 18-36.

Beven, K. J. (2009), Environmental Modelling: An Uncertain Future? *CRC press: London.*

Beven, K. J., and Binley, A. M. (1992), The future of distributed models: model calibration and uncertainty prediction. *Hydrological Processes,* 6(3), 279–298.

Beven, K. J., and Freer, J. (2001), Equifinality, data assimilation, and uncertainty estimation in mechanistic modelling of complex environmental systems. *Journal of Hydrology,* 249, 11-29.

Blazkova, S., and Beven, K. (2009), A limits of acceptability approach to model evaluation and uncertainty estimation in flood frequency estimation by continuous simulation: Skalka catchment, Czech Republic. *Water Resources Research,* 45, W00B16.

Brandimarte, L., and Di Baldassarre, G. (2012), Uncertainty in design flood profiles derived by hydraulic modelling. *Hydrology Research,* 43(6), 753-761. doi:10.2166/nh.2011.086, 2012.

Brandimarte, L., and Woldeyes, M. K. (2013), Uncertainty in the estimation of backwater effects at bridge crossings. *Hydrological Processes,* 27, 1292-1300. doi: 10.1002/hyp.9350.

Brocca, L., Melone, F., Moramarco, T., and Singh, V. P. (2009), Assimilation of observed soil moisture data in storm rainfall-runoff modelling. *Journal of Hydrologic Engineering,* 14(2), 153–165.

Caddis, B., Nielsen, C., Hong, W., Tahir, P. A., and Teo, F. Y. (2012), Guidelines for floodplain development – a Malaysian case study. *International Journal of River Basin Management,* 10 (2), 161-170.

Casas, A., Benito, G., Thorndycraft, V. R., and Rico, M. (2006), The topographic data source of digital terrain models as a key element in the accuracy of hydraulic flood modelling. *Earth Surface Processes and Landforms,* 31: 444–456. doi: 10.1002/esp.1278.

Castellarin, A,. Di Baldassarre, G., and Brath, A. (2011), Floodplain management strategies for flood attenuation in the River Po. *River Research and Applications*, 27, 1037-1047.

Castellarin, A., Di Baldassarre, G., Bates, P. D., and Brath, A. (2009), Optimal cross-section spacing in Preissmann scheme 1-D hydrodynamic models. *Journal of Hydraulic Engineering-ASCE*, 135(2), 96–105. doi: 10.1061/(ASCE)0733-9429(2009)135:2(96).

Charlton, M. E., Large, A. R. G., and Fuller, I. C. (2003), Application of airborne LiDAR in river environments: The River Coquet, Northumberland, UK. *Earth Surface Processes and Landforms*, 28(3), 299-306.

Charrier, R., and Li, Y. (2012), Assessing resolution and source effects of digital elevation models on automated floodplain delineation: A case study from the Camp Creek Watershed, Missouri. *Applied Geography*, 34, 38-46.

Chow, V. T. (1959), Open-Channel Hydraulics. *McGraw-Hill: New York.*

Cobby, D. M., and Mason, D. C. (1999), Image processing of airborne scanning laser altimetry for improved river flood modelling. *ISPRS Journal of Photogrammetry and Remote Sensing,* 56, 121-138.

Cobby, D. M., Mason, D. C., Horritt, M. S., and Bates, P. D. (2003), Two-dimensional hydraulic flood modelling using a finite-element mesh decomposed according to vegetation and topographic features derived from airborne scanning laser altimetry. *Hydrological Processes*, 17(10), 1979–2000.

Cook, A., and Merwade, V. (2009), Effect of topographic data, geometric configuration and modelling approach on flood inundation mapping. *Journal of Hydrology*, 377 (1-2), 131-142.

Crispino, G., Gisonni, C., and Iervolino, M. (2014), Flood hazard assessment: comparison of 1-D and 2-D hydraulic models. *International Journal River Basin Management*, 1-14.

Cunge, J. A. (1975), Two dimensional modelling of flood plains. Chapter 17 in Unsteady Flow in Open Channels (ed. by K. Mahmood and V. Yevjevich), *Water Resources Publications, Colorado.*

de Almeida, G. A. M., Bates, P., Freer, J. E., and Souvignet, M. (2012), Improving the stability of a simple formulation of the shallow water equations for 2-D flood modeling, *Water Resources Research*, 48, W05528, doi:10.1029/2011WR011570.

de Moel, H., van Alphen, J., and Aerts, J. C. J. H. (2009), Flood maps in Europe – methods, availability and use. *Natural Hazards and Earth System Sciences*, 9, 289–301.

Di Baldassarre, G., and Claps, P. (2011), A hydraulic study on the applicability of flood rating curves. *Hydrology Research*, 42 (1), 10-19, doi: 10.2166/nh.2010.098.

Di Baldassarre, G., and Montanari, A. (2009), Uncertainty in river discharge observations: A quantitative analysis. *Hydrology and Earth System Sciences,* 13, 913–921.

Di Baldassarre, G., and Uhlenbrook, S. (2012), Is the current flood of data enough? A treatise on research needs for the improvement of flood modelling. *Hydrological Processes*, 26, 153–158. DOI: 10.1002/hyp.8226.

Di Baldassarre, G., Castellarin, A., and Brath, A. (2009), Analysis of the effects of levee heightening on flood propagation: example of the River Po, Italy. *Hydrological Sciences Journal*, 54(6), 1007–1017.

Di Baldassarre, G., Castellarin, A., Montanari, A., and Brath, A. (2009), Probability weighted hazard maps for comparing different flood risk management strategies: a case study. *Natural Hazards,* 50, 479-496.

Di Baldassarre, G., Laio, F., and Montanari, A. (2012), Effect of observation errors on the uncertainty of design floods. *Physics and Chemistry of the Earth, Parts A/B/C,* 42-44, 85-90.

Di Baldassarre, G., Montanari, A., Lins, H., Koutsoyiannis, D., Brandimarte, L., and Blöschl, G. (2010), Flood fatalities in Africa: from diagnosis to mitigation. *Geophysical Research Letters,* 37, L22402, doi:10.1029/2010GL045467.

Di Baldassarre, G., Schumann G., Bates, P. D., Freer, J. E., and Beven, K. J. (2010), Flood-plain mapping: a critical discussion of deterministic and probabilistic approaches. *Hydrological Sciences Journal,* 55, 364-376.

DID (Department of Irrigation and Drainage) (2003). National Register River Basins, Volume 2: Updating condition of flooding in Malaysia, Malaysia.

DID (Department of Irrigation and Drainage, Malaysia) (2009), Master plan study on flood mitigation for Johor River basin, Malaysia.

Domeneghetti, A., Vorogushyn, S., Castellarin, A., Merz, B., and Brath, A. (2013), Probabilistic flood hazard mapping: effects of uncertain boundary conditions. *Hydrology and Earth System Sciences,* 17(8), 3127–3140.

Dottori, F., Di Baldassarre, G., and Todini, E. (2013), Detailed data is welcome, but with a pinch of salt: accuracy, precision, and uncertainty in flood inundation modeling. *Water Resources Research,* 49(9), 6079–6085.

EM-DAT (Emergency Events Database) (2012), International Disaster Database from http://www.em-dat.net. Universitè Catholique de Louvain, Brussels, retrieved September 2012.

Emergency Events Database (EM-DAT) (2012), International Disaster Database from http://www.em-dat.net. Universitè Catholique de Louvain, Brussels, retrieved September 2012.

EU (2007), Directive 2007/60/EC of the Parliament and the Council of 23 October 2007 on the assessment and management of flood risks. L288/27–L288/34. http://eur-lex.europa.eu/en/index.htm.

Evans T. A. (1998), GIS Data Exchange for the Hydrologic Engineering Center's Hydraulic and Hydrologic Models, *International Water Resources Engineering Conference Proceedings, ASCE, 786-789.*

Farr, T. G., Rosen, P. A., Caro, E., Crippen, R., Duren, R., Hensley, S., Kobrick, M., Paller, M., Rodriguez, E., Roth, L., Seal, D., Shaffer, S., Shimada, J., Umland, J., Werner, M., Oskin, M., Burbank, D., and Alsdorf, D. (2007), The shuttle radar topography mission. *Reviews of Geophysics*, 45, RG2004, doi:10.1029/2005RG000183.

FEMA (Federal Emergency Management Agency) (2008). Guide to emergency management and related terms, definitions, concepts, acronyms, organisations, programs, guidance, executive orders and legislation [online]. Washington: FEMA. Available from: http://www.training.fema.gov/emiweb/edu/docs/terms%20and%20definitions/Terms%20and%20Definitions.pdf [Accessed February 2015].

Fewtrell, T. J., Bated, P. D., Horritt, M., and Hunter, N. M. (2008), Evaluating the effect of scale in flood inundation modelling in urban environments. *Hydrological Processes*, 22, 5107-5118.

Fewtrell, T. J., Neal, J. C., Bates, P. D., and Harrison, P. J. (2011), Geometric and structural river channel complexity and the prediction of urban inundation. *Hydrological Processes*, 25, 3173-3186.

Franques, J. T., and Yannitell D. W. (1974), Two-Dimensional analysis of backwater at bridges. *Journal of the Hydraulics Division*, 100 (3), 379-392.

French, J. R. (2003), Airborne LiDAR in support of geomorphological and hydraulic modelling. *Earth Surface Processes and Landforms*, 28(3), 321-335.

Gichamo, T. Z., Popescu, I., Jonoski, A., and Solomatine, D. (2012), River cross-section extraction from the ASTER global DEM for flood modelling. *Environmental Modelling & Software*, 31, 37-46. DOI:10.1016/j.envsoft.2011.12.003.

Hailea, A. T., and Rientjesb, T. H. M. (2005), Effects of LIDAR DEM resolution in flood modelling: A model sentitivity study for the city of Tegucigalpa, Honduras. ISPRS Workshop Laser scanning 2005, International Society for Photogrammetry and Remote Sensing (ISPRS), Beijing.

Hodgson, M. E., Jensen, J. R., Schmidt, L., Schill, S., and Davis, B. (2003), An evaluation of LIDAR and IFSAR derived digital elevation models in leaf-on conditions with USGS Level 1 and Level 2 DEMs. *Remote Sensing of Environment*, 84(2), 295-308.

Hooijer, A., Klijn, F., Pedroli, G. B. M., and Van Os, A. G. (2004), Towards sustainable flood risk management in the Rhine and Meuse river basins: synopsis of the findings of IRMA-SPONGE. *River Research and Applications*, 20, 343–357.

Horrit, M. S., and Bates, P. D. (2001), Effects of spatial resolution on a raster based model of flood flow. *Journal of Hydrology*, 253, 239-249.

Horritt, M. S. (2005), Parameterisation, validation and uncertainty analysis of CFD models of fluvial and flood hydraulics in the natural environment. Chapter 9,

Computational Fluid Dynamics Applications in Environmental Hydraulics, P. D. Bates, S. N. Lane and R. I. Ferguson, eds. *Wiley, Chichester, UK, 193–213.*

Horritt, M. S. (2006), A methodology for the validation of uncertain flood inundation models. *Journal of Hydrology,* 326, 153-165.

Horritt, M. S., and Bates, P. D. (2002), Evaluation of 1-D and 2-D numerical models for predicting river flood inundation. *Journal of Hydrology*, 268 (1–4), 87–99.

Horritt, M. S., Bates, P. D., and Mattinson, M. J. (2006), Effects of mesh resolution and topographic representation in 2-D finite volume models of shallow water fluvial flow. *Journal of Hydrology*, 329(1–2), 306–314.

Horritt, M. S., Di Baldassarre, G., Bates, P. D., and Brath, A. (2007), Comparing the performance of 2-D finite element and finite volume models of floodplain inundation using airborne SAR imagery. *Hydrological Processes*, 21, 2745-2759.

Hossain, F., Anagnostou, E. N., Dinku, T., and Borga, M. (2004), Hydrological model sensitivity to parameter and radar rainfall estimation uncertainty. *Hydrological Processes*, 18(17), 3277–3291.

Hunter, N. M., Bates, P. D., Horritt, M. S., De Roo, A. P. J., and Werner, M. G. F. (2005), Utility of different data types for calibrating flood inundation models within GLUE framework. *Hydrology and Earth System Sciences,* 9(4), 412-430.

Hydrologic Engineering Center (2001), *Hydraulic Reference Manual.* US Army Corps of Engineers, Davis, California, USA.

ISDR (International Strategy for Disaster Reduction) (2009), UNISDR Terminology on Disaster Risk Reduction, http://www.unisdr.org/eng/library/lib-terminology-eng.htm (last visit November 2010).

Jung, Y., and Merwade, V. (2012), Uncertainty quantification in flood inundation mapping using generalized likelihood uncertainty estimate and sensitivity analysis. *Journal of Hydrologic Engineering,* 17, 507-520.

Jung, Y., Merwade, V., Kim, S., Kang, N., Kim, Y., Lee, K., Kim, G., and Kim, H. S. (2014), Sensitivity of subjective decisions in the GLUE methodology for quantifying the uncertainty in the flood inundation map for Seymour Reach in Indiana, USA. *Water,* 6, 2104-2126.

Jung, Y., Merwade, V., Yeo, K., Shin, Y., and Lee, S. O. (2013), An approach using 1-D hydraulic model, Landsat imaging and generalized likelihood uncertainty estimation for an approximation of flood discharge. *Water,* 5, 1598-1621.

Klijn, F. (2009), Flood risk assessment and flood risk management: An introduction and guidance based on experience and findings of FLOODsite, *FLOODsite-Report T29-09-01*

Krueger, T., Freer, J., Quinton, J. N., Macleod, C. J. A., Bilotta, G. S., Brazier, R. E., Butler, P., and Haygarth, P. M. (2010), Ensemble evaluation of hydrological model hypotheses. *Water Resources Research,* 46(7), W07516. doi: 10.1029/2009WR007845.

Krzysztofowicz, R. (1999), Bayesian theory of probabilistic forecasting via deterministic hydrologic model. *Water Resources Research,* 35(9), 2739–275.

Krzysztofowicz, R. (2002), Probabilistic flood forecast: Bounds and approximations. *Journal of Hydrology,* 268(1–4), 41–55.

Kuczera, G. (1996), Correlated rating curve error in flood frequency inference. *Water Resources Research,* 32(7), 2119-2127.

Li, J., and Wong, W. S. D. (2010), Effects of DEM sources on hydrologic applications. *Computer, Environment and Urban Systems,* 34, 251-261.

Liu, Y. B., Batelaan, O., De Smedt, F., Poórovó, J., and Velcicka, L. (2005), Automated calibration applied to a GIS based flood simulation model using PEST. In: van Alphen J, van Beek E, Taal M (eds) Floods, from defense to management, *Taylor and Francis Group, London, 317–326*.

Mantovan, P., and Todini, E. (2006), Hydrological forecasting uncertainty assessment: incoherence of the GLUE methodology. *Journal of Hydrology, 330(1-2)*, 368-381.

Mark, O., Weesakul, S., Apirumanekul, C., Aroonnet, S. B., and Djordjevic, S. (2004), Potential and limitations of 1-D modelling of urban flooding. *Journal of Hydrology*, 299(3-4), 284–299.

Marks, K., and Bates, P. D. (2000), Integration of high resolution topographic data with floodplain flow models. *Hydrological Processes*, 14 (11-12), 2109-2122.

Martini, F., and Loat, R. (2007), Handbook on good practices for flood mapping in Europe. Paris/Bern: European exchange circle on flood mapping (EXCIMAP), 2007. Available at http://ec.europa.eu/environment/water/flood_risk/flood_atlas/ (last accessed on November 2010).

Maskey, S., Guinot, V., and Price, R. K. (2004), Treatment of precipitation uncertainty in rainfall-runoff modelling: A fuzzy set approach. *Advances in Water Resources*, 27(9), 889–898.

Merwade, V. M., Olivera, F., Arabi, M., and Edleman, S. (2008), Uncertainty in flood inundation mapping – current issues and future directions. *Journal of Hydrologic Engineering*, 13(7), 608–620.

Merwade, V., Cook, A., and Coonrod, J. (2008), GIS techniques for creating river terrain models for hydrodynamic modeling and flood inundation mapping. *Environmental Modelling & Software*, 23 (10-11), 1300-1311.

Merz, B., and Thieken, A. H. (2005), Separating natural and epistemic uncertainty in flood frequency analysis. *Journal of Hydrology*, 309 (1-4), 114–132.

Merz, B., Thieken, A. H., and Gocht, M. (2007), Flood risk mapping at the local scale: concepts and challenges. In: S. Begum, M. J. F. Stive & J.W. Hall, eds. Flood risk management in Europe. Innovation in policy and practice. *Advances in Natural and Technological Hazards Research*, 25, 231–251.

Methods, H., Dyhouse, G., Hatchett, J., and Benn, J. (2007), Floodplain modeling using HEC-RAS, Bentley Institute Press, USA.

Montanari, A. (2005), Large sample behaviors of the generalized likelihood uncertainty estimation (GLUE) in assessing the uncertainty of rainfall-runoff simulations. *Water Resources Research*, 41(8), W08406. doi: 10.1029/2004WR003826.

Moya Quiroga, V., Popescu, I., Solomatine, D.P., and Bociort, L. (2013), Cloud and cluster computing in uncertainty analysis of integrated flood models. *Journal of Hydroinformatics*, 15(1), 55-69, doi:10.2166/hydro.2012.017.

Mukolwe, M. M., Di Baldassarre, G., Werner, M., and Solomatine, D. P. (2013), Flood modelling: parameterisation and inflow uncertainty. *Proceedings of the Institution of Civil Engineers-Water Management*, 167(WM1), 51-60.

Neal, J., Villanueva, I., Wright, N., Willis, T., Fewtrell, T., and Bates, P. (2012), How much physical complexity is needed to model flood inundation?, *Hydrological Processes*, 26 (15), 2264–2282.

Noman, N., Nelson, E., and Zundel, A. (2001), Review of automated floodplain delineation from digital terrain models. *Journal of Water Resources Planning and Management*, 127(6), 394-402.

Paiva, R. C. D., Collischonn, E., and Tucci, C. E. M. (2011), Large scale hydrologic and hydrodynamic modeling using limited data and a GIS based approach. *Journal of Hydrology,* 406, 170-181. DOI:10.1016/j.jhydrol.2011.06.007.

Pappenberger, F., Beven, K. J., Horritt, M., and Blazkova, S. (2005), Uncertainty in the calibration of effective roughness parameters in HEC-RAS using inundation and downstream level observations. *Journal of Hydrology,* 302 (1-4), 46-69.

Pappenberger, F., Beven, K., Ratto, M., and Matgen, P. (2008), Multi-method global sensitivity analysis of flood inundation models. *Advances in Water Resources,* 31, 1-14.

Pappenberger, F., Matgen, P., Beven, K., Henry, J., Pfister, L., and Fraipont de, P. (2006), Influence of uncertain boundary conditions and model structure on flood inundation predictions. *Advances in Water Resources,* 29(10), 1430–1449.

Patro, S., Chatterjee, C., Mohanty, S., Singh, R., and Raghuwansi, N. S. (2009), Flood inundation modelling using MIKE FLOOD and remote sensing data. *Journal of the Indian Society of Remote Sensing,* 37 (1), 107-118.

Poretti, I., and De Amicis, M. (2011), An approach for flood hazard modelling and mapping in the medium Valtellina. *Natural Hazards and Earth System Sciences,* 11, 1141-1151.

Prinos, P., Kortenhaus, A., Swerpel, B., Jimenez, J. A., and Samuels, P. (2008), Review of Flood Hazard Mapping. *FLOODsite, Report-T03-07-01, 42-46.*

Pullar, D., and Springer, D. (2000), Towards integrating GIS and catchment models. *Environmental Modelling & Software,* 15 (5), 451–459.

Purvis, M. J., Bates, P. D., and Hayes, C. M. (2008), A probabilistic methodology to estimate future coastal flood risk due to sea level rise. *Coastal Engineering,* 55(12), 1062–1073.

Rabus, B., Eineder, M., Roth, A., and Bamler, R. (2003), The shuttle radar topography mission - a new class of digital elevation models acquired by spaceborne radar. *ISPRS Journal of Photogrammetry and Remote Sensing*, 57(4), 241–262. DOI:10.1016/S0924-2716(02)00124-7.

Refsgaard, J. C., van der Sluijs, J. P., Højberg, A. L., and Vanrolleghem, P. A. (2007), Uncertainty in the environmental modelling process-A framework and guidance. *Environmental Modelling & Software*, 22(11), 1543-1556.

Renard, B., Kavetski, D., Kuczera, G., Thyer, M., and Franks, S. W. (2010), Understanding predictive uncertainty in hydrologic modeling: the challenge of identifying input and structural errors. *Water Resources Research*, 46, W05521. DOI: 10.1029/2009WR008328.

Robbins, C., and Phipps, S. P. (1996), GIS/Water resources tools for performing floodplain management modelling analysis. *In: Proceedings of AWRA (American Water Resources Association) Symposium on GIS and Water Resources, Fort Lauderdale, FL, (http://www.awra.org/proceedings/gis32/woolprt3/index.html)*.

Rodríguez, E., Morris, C. S., Belz, J. E., Chapin, E. C., Martin, J. M., Daffer, W., and Hensley, S. (2005), *An assessment of the SRTM topographic products, Technical Report JPL D-31639*. Jet Propulsion Laboratory, Pasadena, California. 143 pp., http://www2.jpl. nasa.gov/srtm/srtmBibliography.html (Accessed December 16, 2013),

Romanowicz, R., and Beven, K. J. (1998), Dynamic real-time prediction of flood inundation probabilities. *Hydrological Sciences Journal*, 43(2), 181–196.

Samuels, P. G. (1990), Cross-section location in 1-D models. Proc. International Conference on River Flood Hydraulics, White W.R., John Wiley: Chichester; 339-350.

Samuels, P. G. (1995), Uncertainty in flood level prediction. Proceedings XXVI Biannual Congress of the IAHR. HYDRA2000, Thomas Telford, London.

Samuels, P., Gouldby, B., Klijn, F., Messner, F., van Os, Ad., Sayers, P., Schanze, J., and Udale-Clarke, H. (2009), *Language of Risk: Project Definitions (Second edition)*, Floodsite, T32-04-01, 16.

Schumann, G., Di Baldassarre, G., Alsdorf, D., and Bates, P. D. (2010), Near real-time flood wave approximation on large rivers from space: application to the River Po, Northern Italy. *Water Resources Research*, 46, W05601, DOI:10.1029/2008WR007672.

Schumann, G., Matgen, P., Cutler, M. E. J., Black, A., Hoffmann, L., and Pfister, L. (2008), Comparison of remotely sensed water stages from LiDAR, topographic contours and SRTM. *ISPRS Journal of Photogrammetry and Remote Sensing*, 63, 283 – 296.

Shafie, A., (2009), Extreme flood event: A case study on floods of 2006 and 2007 in Johor, Malaysia. In partial fulfilment of the requirements for Degree Master of Science, Colorado State University Fort Collins, Colorado.

Shrestha, D. L., Kayastha, N., and Solomatine, D. P. (2009), A novel approach to parameter uncertainty analysis of hydrological models using neural networks. *Hydrology and Earth System Sciences*, 13, 1235–1248.

Stedinger, J. R., Vogel, R. M., Lee, S.U., and Batchelder, R. (2008), Appraisal of the generalized likelihood uncertainty estimation (GLUE) method. *Water Resources Research*, 44: W00B06. doi:10.1029/2008WR006822.

Sun, G., Ranson, K. J., Kharuk, V. I., and Kovacs, K. (2003), Validation of surface height from shuttle radar topography mission using shuttle laser altimetry. *Remote Sensing of Environment*, 88(4), 401–411. DOI:10.1016/j.rse.2003.09.001.

Tarekegn, T. H., Haile, A. T., Rientjes, T., Reggiani, P., and Alkema, D. (2010), Assessment of an ASTER generated DEM for 2-D flood modelling. *International Journal of Applied Earth Observation and Geoinformation*, 12, 457–465. DOI:10.1016/j.jag.2010.05.007.

Tayefi, V., Lane, S. N., Hardy, R. J., and Yu, D. (2007), A comparison of one- and two-dimensional approaches to modelling flood inundation over complex upland floodplains. *Hydrological Processes*, 21 (23), 3190–3202.

Trigg, M. A., Wilson, M. D., Bates, P. D., Horritt, M. S., Alsdorf, D. E., Forsberg B. R., and Vega, M. C. (2009), Amazon flood wave hydraulics. *Journal of Hydrology*, 374, 92-105.

UNISDR (United Nations Office for Disaster Risk Reduction) 2002. Guidelines for reducing flood losses. United Nations.

van Alphen, J., and Passchier, R. (2007), Atlas of Flood Maps, examples from 19 European countries, USA and Japan. The Hague, the Netherlands: Ministry of Transport, Public Works and Water Management. Available at http://ec.europa.eu/environment/water/flood_risk/flood_atlas/ (last accessed on November 2010).

van Alphen, J., Martini, F., Loat, R., Slomp, R., and Passchier, R. (2009), Flood risk mapping in Europe, experiences and best practices. *Journal of Flood Risk Management*, 2, 285-292.

Vázquez, R. F., Beven, K., and Feyen, J. (2009), GLUE based assessment on the overall predictions of a MIKE SHE application. *Water Resources Research*, 23, 1325-1349. doi: 10.1007/s11269-008-9329-6.

Walker, W. E., Harremoës, P., Rotmans, J., van der Sluijs, J. P., van Asselt, M. B. A., Janssen, P., and Krayer von Krauss, M. P. (2003), Defining uncertainty a conceptual basis for uncertainty management in model-based decision support. *Integrated Assessment*, 4(1), 5-17.

Wang, W., Yang, X., and Yao, T. (2011), Evaluation of ASTER GDEM and SRTM and their suitability in hydraulic modelling of a glacial lake outburst flood in southeast Tibet. *Hydrological Processes*, 26, 213-225. DOI: 10.1002/hyp.8127.

Ward, R. C. (1978), Floods: A Geographical Perspective. London: Macmillan.

Werner, M. J. F. (2001), Impact of grid size in GIS based flood extent mapping using a 1-D flow model. *Physics and Chemistry of the Earth, Part B: Hydrology, Oceans and Atmosphere*, 26(7–8), 517–522.

Westaway, R. M., Lane, S. N., and Hicks, D. M. (2003), Remote survey of large-scale braided, gravel-bed rivers using digital photogrammetry and image analysis. *International Journal of Remote Sensing*, 24(4), 795-815.

Wilson, M. D., and Atkinson, P. M. (2005), The use of elevation data in flood inundation modelling: a comparison of ERS interferometric SAR and combined contour and differential GPS data. *International Journal of River Basin Management*, 3(1), 3-20. DOI: 10.1080/15715124.2005.9635241.

Wisner, B., Blakie, P., Cannon, T., and Davis, I. (2003), At Risk: Natural hazards, people's vulnerability, and disasters, *Second Edition, Routledge, London and New York, 45.*

Yan, K., Di Baldassarre, G., and Solomatine, D. P. (2013), Exploring the potential of SRTM topographic data for flood inundation modelling under uncertainty. *Journal of Hydroinformatics*, 15(3), 849-861. doi:10.2166/hydro.2013.137.

Yatheendradas, S., Wagener, T., Gupta, H., Unkrich, C., Goodrich, D., Schaffner, M., and Stewart, S. (2008), Understanding uncertainty in distributed flash-flood forecasting for semi-arid regions. *Water Resources Research*, 44, W05S19.

Yu, D., and Lane, S. N. (2006), Urban fluvial flood modelling using a two-dimensional diffusion-wave treatment, part 1: mesh resolution effects. *Hydrological Processes*, 20, 1541-1565.

Acknowledgement

I would like to extend my deepest appreciation to all of those who have contributed either directly or indirectly to my research.

First of all, I would like to expressed my gratitude to Public Service Department (PSD), Malaysia and Department of Irrigation and Drainage (DID), Malaysia for their financial support and study leave throughout my studies. Also, I would like to thank DID for providing the data and giving me the permission to publish this work.

Next, I would like to extend my gratefulness to my promoter Prof. dr. Dimitri P. Solomatine, who I valued the great in-depth discussions, suggestions, critical mindset and ideas that you have especially on the technicality of this research.

My wholehearted appreciation, gratefulness and humble respect go to my co-Promotor/supervisor, Prof. dr. Giuliano Di Baldassarre from Uppsala University, Sweden for his constant support, for trusting my capability to do independent research, for sharing his experience in paper writing and providing valuable and constructive suggestions to complete my thesis.

My acknowledgements are further conveyed to all committee members for their time in reviewing the content of this thesis. My acknowledgments are also dedicated to colleagues in the Hydroinformatics core group: Micah, Isnaeni, Kun Yan, Mario, Juan Carlos and Nagendra for sharing good inputs in this doctoral research, as well as their great sense of humour. I also would like to thank my past and present fellow research comrades: Fiona, Shahrizal, Yuli, Lin, Siek, Girma, Yared, Zahrah, Neiler, and others for their friendship and also insightful discussion – not only scientific but other matters as well. May our friendship be blessed and will last forever. Not forgotten all administrative staff in IHE, Gerda, Jos, Anique, Jolanda, Silvia, Marielle thank you very much for your help.

Special appreciation to Abang Zahari and Kak Noriah, Haji Nordin, Junadi, Faiz, Asrul, arwah or Zaidi and their families, as well as Tio and Yuli. Thank you very much for being part of my family in Netherlands. And not to forget the neighbours in "Bogardhoek" who always make me and my family feel comfortable living in the neighbourhood.

I would also like to express my sincerest gratefulness to many of my friends in Malaysia either in DID or with other organization for their never ending support and true friendship namely: Azad, Anas, Ir Mazura, Marzuki, Nor Azlan, Zain, Abu Zaman, Hafiz K., Wanidi, Dato' Zainor, Dato' Qahar, Dato' Norhisham, Ir. Abdullah, Ir. Sabri, Ir. Ahmad, Ir. Syukri, Ir. Hairi, Ir. Norizan, Ir. Paridah, Aisyah, Zura, Azah, Nizam G., Ir. Raja, Ijat, Salim, Sara, Iena, Fida, Hilman, Noor, Azie, Zaini, Wan, Zul, Aril, Faizal, Maslina, Zana, Duan, Nat, Fizah, Muz, Taff, Mie and Hawa.

Last but not least, no word can describe my indebtedness to my wife, Nor Zalina Ahmad Ahtar together with my beautiful and wonderful family: Nur Aisyah, Nur Alya, Nur Azhar and Nur Azam for their understanding, support, patience, untiring commitments and for everything. Heartiest appreciation to my late parents (in-memory of Halijah Md. Yasin and Md. Ali Ismail - may their gentle soul rest in peace), parents-in-law (Hj. Ahmad Ahtar & Hjh. Mazidah) as well as my brothers (Hamzah, Nizam & Riduan) and sisters (Sharifah & Hamidah) for their support. And not to forget, my thanks to all Malaysian tax-payer: "Without your tax contribution I would not be where I am today".

There are some I may have missed, but they are no less appreciated! Many thanks to all.

"Tiada kekecewaan pada diri ini di dalam menyempurnakan pengajian melainkan tiadanya arwah bonda dan arwah ayahanda untuk bersama meraikannya"

Anuar Md. Ali

Delft, March 2018

About the Author

Anuar Md. Ali was born on 6 February, 1973. He was raised, grown and received early education in Kota Tinggi, Johor, Malaysia. In 1998, he obtained his Degree in Civil Engineering from Universiti Teknologi Malaysia. He later obtained his Master of Science in Water Engineering from Universiti Putra Malaysia in 2004. He started his career in 1998 as a design engineer with a consulting engineering firm, Syed Muhammad, Hooi and Binnie. He joined the Government of Malaysia in 2004 and currently served as an engineer at the Department of Irrigation and Drainage (DID), Malaysia.

In 2009, he was awarded the scholarship by the Government of Malaysia to further his PhD study in UNESCO-IHE, Netherlands.

List of Publications

Journals

Md Ali, A., Di Baldassarre, G., and Solomatine, D. P. (2015). Testing different crosssection spacing in 1-D hydraulic modelling: A case study on Johor River, Malaysia. *Hydrological Sciences Journal.* 60 (2), 351-360, doi: 10.1080/02626667.2014.889297. (published)

Md Ali, A., Solomatine, D. P., and Di Baldassarre, G. (2015). Assessing the impact of different sources of topographic data on 1-D hydraulic modelling of floods. *Hydrology and Earth System Sciences.* 19, 1-13, doi: 10.5194/hess-19-1-2015 (published)

Md Ali, A., Solomatine, D. P., and Di Baldassarre, G. (2016). Uncertainty in simulating design flood profiles and inundation maps on the Johor River, Malaysia. *Journal of Hydrology: Regional Studies* (to be submitted)

Proceedings of International Conferences

Md Ali, A., Di Baldassarre, G., and Solomatine, D. P. (2013). The impact of bridges on flood propagation and inundation modelling. In proceeding of 35th IAHR World Congress 2013, Chengdu, China.

Md Ali, A., Di Baldassarre, G., and Solomatine, D. P. (2014). Impact of different source of topographic information on hydraulic modelling of floods: application to the Johor River, Malaysia. In proceeding of 3rd IAHR Europe Congress 2014, Porto, Portugal. ISBN 978-989-96479-2-3.

Md Ali, A., Di Baldassarre, G., and Solomatine, D. P. (2017). Comparison Between Deterministic And Uncertainty Approaches In Simulating Design Flood Profile On The Johor River, Malaysia. In proceeding of 37th IAHR World Congress 2017, Kuala Lumpur, Malaysia.

Other Conference Contributions

Md. Ali, A., Di Baldassarre, G. and Solomatine, D. P. (2011). Flood hazard mapping under uncertainty: Application to Sungai Johor Basin, Malaysia. EGU Leonardo Conference, Bratislava. (Poster presentation)

Md. Ali, A., Di Baldassarre, G. and Solomatine, D. P. (2011). Flood risk mapping under uncertainty: Application to Sungai Johor Basin, Malaysia. UNESCO-IHE Annual Phd Seminar 2011, Netherlands. (Oral presentation)

Md. Ali, A., Di Baldassarre, G. and Solomatine, D. P (2011). Impact of topographic uncertainty on flood inundation modelling. UNESCO-IHE Hydroinformatic Seminar 2011, Netherlands. (Oral presentation)

Md. Ali, A., Di Baldassarre, G. and Solomatine, D. P (2012). Determination of cross section spacing in 1-D hydraulic model. UNESCO-IHE Annual Phd Seminar 2012, Netherlands. (Oral presentation)

Md. Ali, A., Di Baldassarre, G. and Solomatine, D. P (2013). Influence of structures on flood hazard. EGU General Assembly, Geophysical Research Abstracts, EGU 2013. (Poster presentation)

Md. Ali, A., Di Baldassarre, G. and Solomatine, D. P (2013). The impact of bridges on flood propagation and inundation modelling. 35th IAHR World Congress 2013, Chengdu, China. (Oral presentation)

Md Ali, A., Di Baldassarre, G., and Solomatine, D. P. (2014). Impact of different source of topographic information on hydraulic modelling of floods: application to the Johor River, Malaysia. 3rd IAHR Europe Congress 2014, Porto, Portugal. (Oral presentation)

Md Ali, A., Di Baldassarre, G., and Solomatine, D. P. (2017). Comparison Between Deterministic And Uncertainty Approaches In Simulating Design Flood Profile On The Johor River, Malaysia. In proceeding of 37th IAHR World Congress 2017, Kuala Lumpur, Malaysia. (Oral presentation)

Index of Notation and Abbreviations

1-D	One-Dimensional
2-D	Two-Dimensional
ASTER	Advanced Spaceborne Thermal Emission and Reflection Radiometer
CFL	Courant-Friedrich-Levy
CGIAR-CSI	Consortium of International Agricultural Research Centers-Consortium for Spatial Information
DEM	Digital Elevation Model
DGPS	Differential Global Positioning System
DHI	Danish Hydraulic Institute
DID	Department of Irrigation and Drainage, Malaysia
DSMP	Department of Survey and Mapping, Malaysia
DTCP	Department of Town and Country Planning
DTM	Digital Terrain Model
EM-DAT	Emergency Events Database
EU	European Union
EXCIMAP	European Exchange Circle On Flood Mapping
FEMA	Federal Emergency Management Agency
GIS	Geographic Information System
GLUE	Generalized Likelihood Uncertainty Estimation
GPS	Global Positioning System

ICPDR	International Commission for the Protection of the Danube River
ICPR	International Commission of the Protection of the Rhine
ISDR	International Strategy for Disaster Reduction
JICA	Japan International Corporation Agency
J-spacesystems	Japan Space Systems
LiDAR	Light Detection and Ranging
MAE	Mean Absolute Error
METI	Ministry of Economy, Trade and Industry
NASA	National Aeronautics and Space Administration
NGA	National Geospatial-Intelligence Agency
PEST	Parameter Estimation by Sequential Testing
RMSE	Root Mean Square Error
SAR	Synthetic Aperture Radar
SRTM	Shuttle Radar Topography Mission
TIN	Triangulated Irregular Network
UK	United Kingdom
UNISDR	United Nations Office for Disaster Risk Reduction
USD	US Dollar
USGS	United States Geological Survey
WSE	Water Surface Elevation

Printed and bound by CPI Group (UK) Ltd, Croydon, CR0 4YY

22/10/2024

01777647-0009